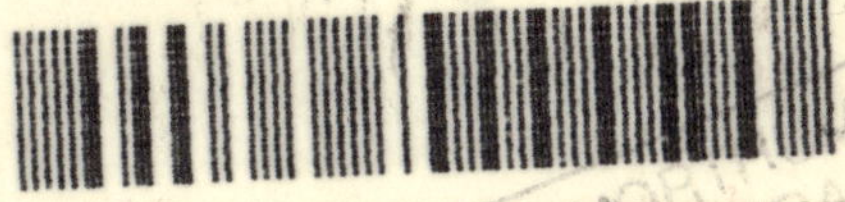

Quality Assurance of Welded Construction

Quality Assurance of Welded Construction

Edited by

N. T. BURGESS

Director, Quality Assurance Services,
Gilbert Commonwealth Engineers and Consultants Ltd,
Twickenham, Middlesex, UK

APPLIED SCIENCE PUBLISHERS
LONDON AND NEW YORK

APPLIED SCIENCE PUBLISHERS LTD
Ripple Road, Barking, Essex, England

Sole Distributor in the USA and Canada
ELSEVIER SCIENCE PUBLISHING CO., INC.
52 Vanderbilt Avenue, New York, NY 10017, USA

British Library Cataloguing in Publication Data

Quality assurance of welded construction.
1. Welding
I. Burgess, N.T.
671.5′2 TS227.2
ISBN 0-85334-184-2

WITH 13 TABLES AND 54 ILLUSTRATIONS

Filmset and printed in Northern Ireland at The Universities Press (Belfast) Ltd.

Preface

The growing application of quality assurance both as a regulatory and contractual requirement and as a management discipline for the modern supplier has had significant impact on welding, the most important of the manufacturing processes.

At the same time, the uses of welding increase daily with the drive towards more economic construction for an ever widening range of industries. Problems with welded equipment still arise from a variety of causes, many of which are dealt with in this book. In the early days of welding both manufacturers and users were tolerant of fabricating and construction difficulties but this is no longer possible as the cost of failure, re-work and inspection increases. Further, the development of welding techniques themselves, the metals that can be joined and the range of thicknesses involved, have contributed more problems. There is a need to minimise at every point the influence of these factors on potential failure and on the avoidance of defects.

Quality assurance has developed as a total control concept without specific relevance to welding or indeed any manufacturing method, and it has demonstrated its value in maintaining and improving quality and safety standards, wherever possible in an economic fashion.

The object of this book is to bring together, it is believed for the first time, the basic principles and techniques of quality assurance in relation to a specific area of industry, tailored to a major construction method. It has, within the confines of one volume, been difficult to decide what to include and what to exclude and since quality assurance can be said to embrace design phase and metallurgical aspects, as well as construction practices, it has only been possible to concentrate on the cardinal issues and on some valuable ideas from the contributors.

Whilst the basic concepts and procedures contained herein are applicable to any welded construction the authors have in general been drawn from the 'heavy' end of industry and therefore examples and case studies referenced relate thereto. This will be very relevant to those industries that typically use pressure vessels, pipework, process plant, bridges, mechanical handling and like structures. As such, welding is most evident in the context of metal arc, inert gas, submerged arc and related methods, although resistance spot welding, for example, amongst other joining techniques, is not discussed. Since both quality assurance and welding principles span international boundaries reference is made where possible to internationally used specifications and practices from several countries.

The authors selected for this work have together and individually a vast experience of the application of welding in many industries, particularly those now grappling with the application of quality assurance. They are authorities in their own right and their backgrounds cover the academic field, research, manufacturing, design and consultancy. The Editor's past spans the energy field including nuclear generation, the process plant industry and, more recently, the offshore oil and gas business. His opening chapter is intended to brief the unwary on some aspects of quality assurance that may not be apparent from the contract, the text books or indeed national standards.

It is generally accepted that up to 80 % of engineering problems are ultimately attributable to the design stage and in Chapter 2 Dr Jubb has provided some well-chosen examples of how welded design must take account of modern thinking and the latest knowledge. The author of the extensive coverage on manufacturing, Mr Gifford, had the benefit of working as a manufacturing welding engineer before becoming a QA manager and his insight into shop floor problems, particularly in the boiler and pressure vessels field, is extensive. Much welded work for the industries that form the basis of this treatise takes place at site, be it nuclear power station, oil refinery or gas pipeline, and to complement the workshop operations Mr Butler has provided direct experience of site practices with a bearing on quality assurance. A sound knowledge of the relevant artefacts is essential equipment for the practising quality assurance engineer and Dr Rogerson has succinctly reviewed the key information and commented on defect significance. This is particularly important in relation to acceptance criteria, which are discussed here and in other chapters. Inspection is the subject of many clauses in the national standards on welded plant and the author of this chapter has therefore limited his contribution to those aspects that have a strong bearing on quality assurance such as human aspects and qualification, referencing the techniques which are relevant.

The widespread use of non-destructive testing (NDE in many countries) has confirmed that when used correctly this is a major tool in assuring the quality of weldments. The emphasis in this chapter by Mr Jessop is not merely on techniques, as this can be studied from the textbooks, but on the limitations of the methods in relation to specific defects, and on the consideration of the scientific principles involved. Finally, since the core of a good QA programme rests on an understanding of what is considered 'good practice' between supplier and client, there is a review of standards and codes related to welding which highlight the strengths and weaknesses to which a QA man should address himself.

This then is a book about quality assurance in welding which indicates what is achievable and necessary rather than merely 'How to make good welds'. As such the contributors hope that readers, whether they be quality engineers seeking a greater understanding of welding, or welding people faced with the needs of quality assurance, will be able to tackle their work more effectively.

N. T. BURGESS

The author of this chapter has therefore limited his contribution to those aspects that have a strong bearing on quality assurance, such as human aspects, and qualifications concerning the techniques which are relevant.

The widespread use of non-destructive testing (NDT) in many countries has confirmed that when used correctly this is a major tool in assessing the quality of products. The emphasis in this chapter by [illegible] is not so much on the techniques, as this can be studied in the textbooks, but on the application of the methods in relation to specific defects and on the qualification of the personnel [illegible] involved. Finally, since the concept of [illegible] [illegible] between an ideal situation and what is [illegible] in practice [illegible] [illegible] there is [illegible] which highlight the [illegible] [illegible] QA [illegible] should be [illegible].

[illegible] applicable when a [illegible] is not necessarily [illegible] make good [illegible] the [illegible] that [illegible], whether [illegible] quality [illegible] and funding of [illegible] with the [illegible] quality assurance will be able [illegible].

N. [illegible]

Contents

Contents

List of Contributors

N. T. BURGESS

Director, Quality Assurance Services, Gilbert Commonwealth Engineers and Consultants (UK) Ltd, Fraser House, London Road, Twickenham, Middlesex TW1 3ST, UK.

B. S. BUTLER

Principal Welding Engineer, British Steel Corporation, Teesside Laboratories, Ladgate Lane, PO Box 74, Middlesbrough, Cleveland TS8 9EG, UK.

A. F. GIFFORD

Chief Metallurgist, Northern Engineering Industries (Mechanical Engineering Ltd), Engineering Development Department, Sinfin Lane, Derby DE2 9GJ, UK.

T. J. JESSOP

Head, NDT Research Section, Engineering Department, The Welding Institute, Abington Hall, Abington, Cambridge CB1 6AL, UK.

J. E. M. JUBB

Head of Inspection, Messrs Sandberg, Harpur House, 62 Harpur Street, Bedford MK40 2RA, UK.

J. H. ROGERSON

Senior Lecturer and Consultant in Welding Technology and Quality Assurance, Department of Materials, Cranfield Institute of Technology, Cranfield, Bedford MK43 0AL, UK.

1

Basic Concepts in Quality Assurance

N. T. Burgess
Gilbert Commonwealth Engineers and Consultants (UK) Ltd, Twickenham, UK

INTRODUCTION

Once the satisfactory design of a product or construction has been evolved, detailed and checked, there is a need to specify quality characteristics against which to produce.

The quality of manufactured products is frequently dependent upon the effectiveness of the manufacturer's control of fabrication, inspection and testing operations. In consequence, manufacturers are responsible for instituting such controls over operation, processes and checking, as are necessary to ensure that their products conform to the specified requirements. Today, manufacturers are also often obliged to provide objective, verifiable evidence that they have carried out all necessary activities. This means that a supplier is expected to supply not only products and services but, in addition, proof that the product has been properly made and tested. A measure of assurance can be gained by (the customer) ensuring that everything necessary has been done to achieve the required integrity of each characteristic of the finished product. Thus, 'quality assurance'!

QUALITY ASSURANCE: DEFINITIONS

The generally accepted definition of quality assurance (QA) and related terms are based on those promulgated by the European Organisation for Quality Control [1].

Quality Assurance—all those planned or systematic actions necessary to provide adequate confidence that a product or service will satisfy given needs.

Comment: quality assurance aims at making sure that quality is what it should be. This includes a continuing evaluation of adequacy and effectiveness with a view to having timely corrective measures and feedback initiated where necessary. For a specific product or service, quality assurance involves the necessary plans and actions to provide confidence through verifications, audits and the evaluation of the quality factors that affect the adequacy of the design, inspection and use of the product or service. Providing assurance may involve producing evidence.

Note: when quality assurance is used in the total system sense, as it normally is when used without a restrictive adjective, it embraces all aspects of quality. When it is used in a more restricted sense for a particular phase or function within a total quality assurance system, the phrase 'quality assurance' is modified by an adjective or used as an adjective to restrict some other operation. For example, 'design quality assurance' or 'supplier quality assurance'.

Quality Control—the operational techniques and the activities which sustain a quality of product or service that will satisfy given needs; also the use of such techniques and activities.

Comment: the aim of quality control (QC) is to provide quality that is satisfactory (e.g. safe, adequate, dependable, and economical). The overall system involves integrating the quality aspects of several related steps including: the proper specification of what is wanted; design of the product or service to meet the requirements; production or installation to meet the full intent of the specification; inspection to determine whether the resulting product or service conforms to the applicable specification; and review of usage to provide for revision of specification. Effective utilisation of these technologies and activities is an essential element in the economic control of quality.

Note: when quality control is used in the total system sense, as it normally is without a restrictive adjective, it is concerned with making quality what it should be. When used in a more restricted sense for a particular phase or function within the total quality control system, the phrase 'quality control' is modified by an adjective for example, 'process quality control', 'manufacturing quality control', 'design quality control', etc. or used as an adjective to restrict other operations to that part which belongs within the quality control system, as, for example, 'quality control inspection', 'quality control testing', etc.

Quality Surveillance—supervision of a contractor's or supplier's quality assurance organisation and methods by the customer, his representative or an independent organisation.

Inspection—the process of measuring, examining, testing, gauging or otherwise comparing the unit with applicable requirements.

These basic definitions go to make up the subject of Quality Engineering—that branch of engineering which deals with the principles and practice of product and service quality assurance and control.

A quality engineer may need to be qualified in some or all of the following aspects:

(1) Development and operation of quality assurance and control systems.
(2) Development and analysis of testing, inspection and sampling procedures.
(3) An understanding of the relationship of human factors and motivation with quality.
(4) Facility with quality cost concepts and techniques.
(5) The knowledge and ability to develop and administer management information including the auditing of quality programmes to permit identification and correction of deficiencies.
(6) The ability to arrange appropriate analyses to determine those operations requiring corrective action.
(7) Application of metrology and statistical methods to the analysis of quality parameters for both control and improvement purposes.

THE REASONS FOR QUALITY ASSURANCE

Quality assurance concepts grew out of quality control, which, in the stage of industrial development after the Second World War, became a necessity in many industries. In the USA the use of statistical techniques, particularly in the continuous production industries, telecommunications, etc. was a prime tool in measuring the performance of processes, men and machines. It was possible to predict the likely level of defects in a given situation and therefore attempt to prevent or reduce them. Quality control, as a management or production discipline became standard practice in the USA and, later, in Japan, although application to 'one off' structures, power stations, oil refineries and the like was much slower. In many countries the development of quality control practices progressed through defence equipment to electronics generally, nuclear plants and conventional power stations and to many critical structures.

The incentive for manufacturers to reduce defects and therefore costs led to extensive development of quality control except perhaps in those industries where the purchaser has traditionally taken some responsibility for quality control. In 'one off' and heavy engineering industries often involving welded construction, customer inspection, or rather witnessing inspection has been the convention and it is claimed by many that this led to minimum quality control efforts by respective manufacturers. This appears to be true when compared with industries supplying retail or consumer outlets where market pressures forced a tighter control of quality aspects of the product. 'Quality control', as a tool for suppliers, therefore preceded 'Quality assurance' for customers. (Quality control has been used since the 1940s as a term to include *all* methods used to control quality including inspection and non-destructive testing (NDT).)

'Quality assurance' on the other hand is a term which has grown in the Western World mainly since the 1960s. The term connotes much broader concepts of quality and reliability achievement and involves action by, and responsibilities on, the purchaser/user. (The Allied Quality Assurance Publications (AQAPs) used by NATO [2] are customer produced requirements that formed the basis for many similar customer related requirements).

In welded construction, many aspects of controlling quality have been developed inherently or instinctively, e.g. the selection and use of correct and proven welding parameters. Indeed, resistance spot and similar automatic methods have always lent themselves to quality control—to the exclusion of individual weld inspection.

The extensive use of welding as the most important constructional method and the rapid developments in welding technology itself are two factors that have inevitably contributed to engineering quality problems. Other factors are the more onerous service required of many welded structures or components, the progressive reduction of safety margins and the use of more economic or conservative design criteria, as well as developments in materials to be welded.

Problems with welding, both during manufacture and in-service (by way of failure) have led, over the post-war period, to a tremendous growth in weld inspection (often on a 100 % basis) by customers as well as manufacturers, as an obvious, although sometimes misguided, defence. Third party inspection (by the State, an approved body, or by inspection organisations) has flourished in most countries and in direct relationship to the increase in use of welding. This is especially noticed in the heavy engineering sector, power plants, oil refineries, ship building and so on, particularly where pressure plant and other potentially dangerous equipment is being constructed. Unfortunately much of this effort has been ineffective, indeed, counter productive and throughout this book are indications as to how this situation could be improved.

Admiral Rickover in his now famous 1962 address to the 44th Annual Metal Congress in New York [3] said—'The price of progress is the acceptance of more exacting standards of performance and relinquishment of familiar habits and conventions rendered obsolete because they no longer meet the new standards. To move but one rung up the ladder of civilisation man must surpass himself'. He followed with a catalogue of quality problems besetting the nuclear industry (particularly in nuclear submarines) at that time. The author believes that this paper was a turning point in the move towards quality assurance around the world, subsequently to be taken up in most technological industries and in many countries.

Some of the best publicised failures, like that of the Kings Bridge in Melbourne in 1962 [4] demonstrated a serious lack of quality control. Several pressure vessel failures during tests in the UK and elsewhere during the 1950s and 1960s had direct or indirect relevance to a lack of quality control [5, 6].

Rickover also reported further on serious deficiencies in US manufacturing practices 'During the past few years, hundreds of major conventional components such as pressure vessels and steam generators, have been procured for naval nuclear propulsion plants. Less than 10 % have been delivered on time. 30 % were delivered six months to a year or more later than promised. Even so, re-inspection of these components after delivery showed that over 50 % of them had to be further re-worked in order to meet contract specification requirements'.

And again, 'there are 99 carbon steel welds in one particular nuclear plant steam system. The manufacturer stated that all these welds were radiographed and met specifications. Our own (US Navy) re-evaluation of these welds—using correct procedures and proper X-ray sensitivity—showed however that only 10 % met ASME standards; 35 % had defects definitely in excess of ASME standards and the remaining 55 % had such a rough external surface that the radiographs obtained could not be interpreted with any degree of assurance.'

Serious failures and delays in the UK power programme due to poor quality control blighted the construction programme of the 1960s [7]. Excessive failure levels in welds made by oxyactylene welding, by flash welding, at attachments to boilers, and with defective boiler tubes, as well as large turbine castings, were reported. Since the worst performance came from conventional power plant components rather than nuclear items, the initial corrective programme was in that sector. Major failures associated with welding have been experienced by most industries.

QUALITY ASSURANCE—A POSSIBLE SOLUTION TO ENGINEERING PROBLEMS

Some of the reasons why quality assurance is required by purchasers and plant users are:

(1) The recognition that inspection and tests alone do not *prevent* defects—they may not even prevent them getting into service.
(2) No one can *inspect* quality into a product—it has to be designed or built in.
(3) There are as many failures in inspected plant as there are in uninspected plant (therefore something else needs to be done).
(4) It costs more to make defective welds than good ones.

Control of quality must be planned and organised, just like any other business parameter.

It is clear that, because of the nature of welding operations, inspection (including examination) and monitoring surveillance will be necessary. However, it is fundamental to quality assurance that 'quality is best controlled by those responsible for the product and by those closest to the point of manufacture'. This must mean the supplier himself and not the purchaser. (This aspect is dealt with further in Chapter 6.)

This is particularly important in the 1980s since product liability concerns many countries and many industries. The responsibility for accidents resulting from poor quality design and manufacture is generally placed with the designer or manufacturer [8].

Whose Responsibility is Quality Assurance?

Basically a quality assurance approach by an industry, a user or a supplier requires at least:

(1) A top management policy decision, followed through to line management.
(2) A quality assurance specification or programme identifying the 'management' criteria.
(3) An obligation to evaluate and audit control processes, procedures and instructions as a preliminary to quality control surveillance.

There is no single solution to the avoidance of failures in welded construction; many different factors may be involved, not only within a particular company, but outside, with suppliers, sub-contractors, customers, inspection authorities, etc. What can be achieved is a more disciplined approach to all quality related activities from 'cradle to grave' of a product or component. This includes the salesman who should not give vain promises to the client, the metallurgist who is too often confined to the laboratory and the client who seeks technological impossibilities.

It should also be recognised at the outset that such assurance has to be paid for, either by reduced scrap, by fewer repairs or by less routine inspection (see p. 16). This can be achieved if all parts of an enterprise from chairman or managing director to the packer and despatcher understand the importance of meeting quality objectives. These in turn must be spelled out in sufficient detail at contract stage, in the specification, or on the drawing. These elementary principles of quality assurance are explained in detail in specifications which most developed countries now have, or are developing. In the UK these are BS 4891 A Guide to Quality Assurance, and BS 5750 (Parts 1 and 2) Quality Systems.

The NATO Allied Quality Assurance Publications [2] spell out the same principles which are largely based on the original US Military Specification Mil-Q-9858a. The nuclear power industry has adapted the same principles and more recently the International Standards Organisation, ISO, through its committee ISO TC76, is considering an

TABLE 1

MATRIX OF QUALITY ASSURANCE REQUIREMENTS

Criteria	*Relevant clause*			
	10CFR50 APP B	*Mil-Q 9858A*	*BS 5750 Part 1*	*ANSI N45·2*
Organisation	I	3·1	4·2	3
Quality assurance plan review and approval	II	3·2	4·3, 4·4	2
of plan	PSAR/FSAR	1·2		Not specified
Design control	III	4·1	4.8	4
Procurement document control	IV	5·1 5·2	4·9	5
Instructions, procedures and drawings	V	6·2	4·5, 4·12	6
Document control	VI	4·1	4.9	7
Control of purchased materials	VII	5·1	4.11	8
Identification and control of materials, parts and components	VIII	No traceability	4·11	9
Control of special processes	IX	62		10
Inspection	X	6·3	4·17	11
Test control	XI	6.3	4.17	12
Control of measuring and test equipment	XII	4.2, 4·3 4·4, 4·5	4·10	13
Handling storage and shipping	XIII	6·4	4·18	14
Inspection, test and operating status	XIV	6·7	4·17	15
Non-conforming materials, parts or components	XV	6·5	4·16	16
Corrective action	XVI	3·5	4·7	17
Quality assurance records	XVII	3·4	4·6	18
Audits	XVIII	None	4·3	19

international specification for adoption by general industry [9]. Many standards refer to 18 separate criteria and Table 1 illustrates how these are identified in four such QA standards relating to the defence, nuclear and general industry. Such standards are increasingly being applied to other high hazard or high technology plant, and indeed in general engineering.

Where to Start

Assuming that top management is committed to a quality policy and all of its obligations, certain initial steps would be sensible. A supplier must first determine where quality problems are occurring (using pareto analysis), e.g. design and specification, procurement, material, manufacture, inspection, site construction? Figure 1 illustrates an analysis of problem areas and presents the approximate situation by one UK pressure vessel manufacturer. Discussions with other manufacturers in other industries around the world confirm this general picture.

The significance of the individual must be considered. Welding introduces particular problems, e.g. whilst designers and engineers are professionally qualified and trained, welders and operators rely more

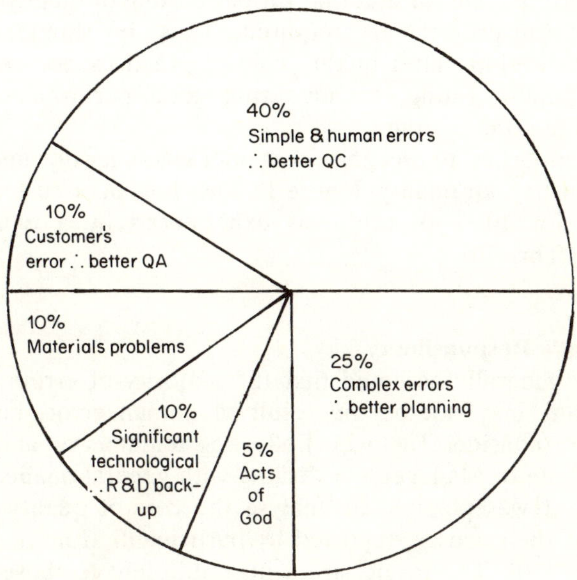

FIG. 1. Typical analysis of quality problem areas.

on personal skills. The results of these efforts are judged by others who may have inadequate training or experience to take the critical decisions required of them (see Chapter 6). Thus:

Skills in successful weldments

Designer	—	Professional training
Welding engineer	—	Professional training
Welder	—	Artisan skills
Welding inspector Non-destructive tester	—	Decision maker

The Purchaser's Responsibility

Most progress in obtaining and operating safe and reliable welded plant made to sound quality assurance principles has been in those industries where the purchaser has been wise enough to recognise the real benefits that accrue from a quality assurance policy. The obligations upon such purchasers are similar to those listed earlier, i.e. top level commitments, a QA requirement or specification reference and the allocation of appropriate resources. Equally important, however, is the need for the purchaser to recognise that improvements in performance are worth paying for and that proper costing of quality aspects of the product (or project) are required. Thus, he should encourage suppliers to develop better quality control practices, for example, by employing vendor rating, by favouring good performance and by employing life cycle costing.

It is also important to recognise that poor customer documentation is a contributor to poor quality. Figure 1 shows how poor customer QA is responsible for 10 % of problems experienced by a typical heavy engineering fabricator.

Management's Responsibility

It is now generally accepted that the majority of errors and faults in engineering may not be the result of human error, but of poor management (consider Table 2). Following seminars organised by the British Institute of Management dealing with general quality problems in industry, it was possible to analyse the cost of quality calamities according to their causes, reported by participants from a wide range of industries [10]. The results are a little different to those in Fig. 1, but Table 2 shows that 'lack of proving' was the most recurrent cause of failure with 'lack of, or inadequate specifications' a major factor

TABLE 2

THE DISTRIBUTION OF QUALITY CALAMITIES ARISING FROM NINE MANAGEMENT SEMINARS IN THE 1960s[a]

Cause	*Number of calamities by the cost of each*						
	£100 000	£50 000	£25 000	£10 000	<£10 000	*All*	%
Human error	$2\frac{1}{2}$	$\frac{1}{2}$	2	6	$3\frac{1}{2}$	$14\frac{1}{2}$	12
Bad inspection methods	$\frac{3}{4}$	$\frac{1}{2}$	$2\frac{1}{2}$	3	$5\frac{1}{2}$	$12\frac{1}{4}$	10
Lack of/or wrong specification	$3\frac{1}{4}$	1	$2\frac{1}{2}$	7	6	$19\frac{3}{4}$	16
Lack of proving (new designs, materials or processes)	$7\frac{1}{4}$	$3\frac{1}{2}$	$7\frac{1}{2}$	$14\frac{1}{3}$	11	$43\frac{1}{2}$	36
Poor planning and co-ordination	$2\frac{1}{4}$	2	$1\frac{1}{4}$	7	$4\frac{1}{2}$	17	14
Unforseeable	$2\frac{1}{2}$	0	$\frac{3}{4}$	2	5	10	8
Other causes	$\frac{1}{2}$	$2\frac{1}{2}$	$1\frac{1}{2}$	0	0	$4\frac{1}{2}$	4
All causes	19 16%	10 8%	18 15%	39 32%	25 29%	121 100%	100

Note. The fractions refer to shared causes.

[a] Courtesy of R. M. Belbin, from a survey [10] into costs and causes of quality calamities quoted by quality practitioners taking part in British Institute of Management seminars in 1968. The causes in rows 3, 4 and 5 together account for 66 % of the total and are considered to be 'management' causes.

along with 'poor planning and co-ordination'—all areas of management responsibility, rather than the operative.

It will be seen that the 'human' error problems in the example are 12 % whereas 'management' deficiencies in planning, specification and proving stages amount to some 66% of cases. If we add 'bad inspection methods' which is often a management responsibility we have 76 % of the causes attributable to management. This general analysis is quite applicable to the case of welding problems. A prime function of management should therefore be 'root cause analysis' and it should be the concern of management to deploy appropriate personnel to this aspect as well as troubleshooting.

The list below indicates some root causes identified by 20 quality assurance managers, both purchasers and suppliers, following a survey conducted by the author in the UK in 1979/80 [11]:

Root causes of poor supplier performance:
Lack of planning
Overbooking of work

Pressure of production over quality
Lack of authority of quality personnel
Lack of clear QC systems
Inadequate in-process/Final inspection
Poor manufacturing equipment
Inadequate briefing on specification/Lack of understanding
Inadequate direction, Poor management, Poor control
Disinterested work force
Poor control of sub-contractors
Lack of day to day QA implementation
Middle management apathy
Poor customer specifications and contractual requirements
Poor QA staff
Belief that quality costs money

The Designer's Responsibility

The inclusion of a full chapter on 'Designing Weldments to Avoid Failure' (Chapter 2), emphasises the importance of design quality assurance. Detailed examination of failures including weld failures, suggests that some 80 % originate at the design stage, or at least the failure could have been avoided by action at the design stage. Design, in this context, includes material selection, and design detail as well as conceptual design and adequacy of specifications. Of particular importance in design assurance is inspectability, i.e. can the weld be inspected both during construction and during later maintenance checks? One of the techniques of QA utilised at the design stage is design review. Design reviews are systematic critical studies of the design or its elements at various stages by specialists not necessarily directly engaged in the design, to provide assurance that the final design will satisfy the specifications. The factors considered at such a review should include:

(a) comprehensiveness of design inputs;
(b) adequacy of assumptions used;
(c) appropriateness of design methods;
(d) correct incorporation of design inputs;
(e) adequacy of the output;
(f) The necessary design input and verification requirements for interfacing organisations are specified in the design documents, or in supporting procedures or instructions;
(g) viability of the project.

Reviews may be performed at the various stages of the design process (e.g. conceptual, intermediate, final) and may be performed by persons drawn from various disciplines (e.g. engineering, design, manufacturing, quality assurance, marketing, financial, personnel, legal) as appropriate to the product or the state of the design. Prior to the design review, a check list of topics to be considered should be established.

In essence, the elementary requirements of QC/QA can be expressed:

QA	QC	Plan what you do	(Written procedures)
		Do what you say	(Practice)
	Record that it has been done		(Records)

THE ECONOMICS OF QUALITY

Too often quality assurance only emerges as a policy issue when governments, the State or traumatic situations force managements to consider that which should be a standard business discipline. Of course, the degree to which QA is applied will vary from country to country or industry to industry. The factors to be considered in following a QA policy may include the following.

For the Supplier

Does the purchaser have a QA requirement or policy?
Will he enforce this in his procurement of plant and equipment?
Does our company want to be in this business?
Will our management support and fund it?

For the Purchaser

Is the industry generally committed to QA?
Do the contractors and suppliers understand QA?
Do I have the necessary detail in my engineering and purchasing specifications?
Am I prepared to accept all of the obligations of a QA programme?

These and other questions must be asked because, as with other business activity, success depends on commitment. Too often quality assurance is accepted as a necessary evil or to obtain a particular contract or to be on an 'approved list', without proper commitment from those whose efforts will determine success or failure.

The economic motive is strong and it is not difficult to see why some purchasers call up quality assurance as a major platform in its defence against:

(1) Costly outages and unplanned shutdown due to premature failure.
(2) Serious manufacturing and construction delays.
(3) Government safety or environmental legislation.
(4) High cost of replacement or repair.
(5) Inefficient suppliers.

Suppliers take up quality assurance (more often quality control) to:

(1) Reduce scrap, repair, rectification and wasted time.
(2) Ensure customer satisfaction.
(3) Meet statutory or contract requirements.
(4) Keep one step ahead of the competition.
(5) Improve efficiency and cost effectiveness.
(6) Reduce customer inspection visits and customer interference.

It must be understood that effective quality assurance is a 'two way' function between supplier and purchaser and that shallow QA systems entered into with inadequate discussion between the contracting parties are often wasted effort costing a great deal of money.

Many major users devote financial and management resources to quality assurance because of the high costs incurred when situations noted earlier occur. The power industry quotes costs up to £40 000 per each day that high merit plant is out of service and other similar process plants incur similar costs (see Table 2).

The defence industry measure failure in other terms. In all industries, welding is a predominant element and it is not surprising that weldments frequently fail, or are involved in failures [3, 4, 5, 6].

Quality Costs

For the supplier quality costs can be divided into:

(a) *Prevention costs*—the cost of any action taken to prevent or reduce defects and failures.
(b) *Appraisal costs*—the cost of assessing the quality achieved.
(c) *Internal failure costs*—the costs of failure to meet quality requirements prior to the transfer of ownership to the customer.

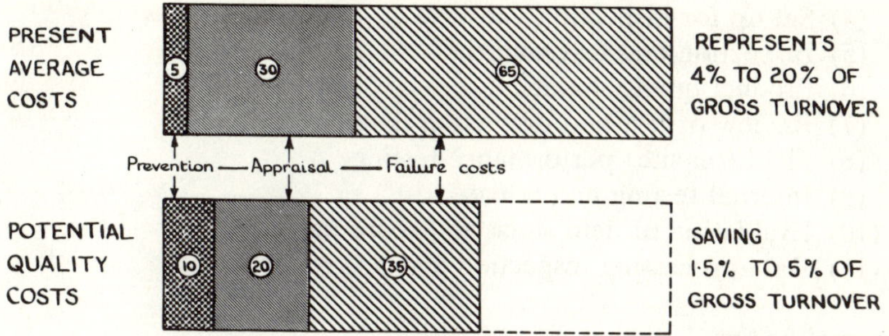

FIG. 2. Investment in prevention reduces failure costs.

(d) *External failure costs*—the costs of failure to meet quality requirements after the transfer of ownership to the customer.

Investment in prevention and appraisal can substantially reduce internal and external failure costs (Fig. 2). Furthermore, reductions in external complaints are important not only to reduce costs but also to maintain customer goodwill.

The British Standard Guide to Quality Related Costs [12] suggests the following groupings for the various items of cost:

Prevention Cost

(1) Quality engineering
  (i) quality control engineering
  (ii) process control engineering
(2) Design and development of quality measurement and control equipment.
(3) Quality planning by other functions.
(4) Calibration and maintenance of production equipment used to evaluate quality.
(5) Maintenance and calibration of test and inspection equipment.
(6) Supplier assurance.
(7) Quality training.
(8) Administration, audit and improvement.

Appraisal Cost

(1) Laboratory acceptance testing.
(2) Inspection and test (including 'goods inward').
(3) In-process inspection.

(4) Set-up for inspection and test.
(5) Inspection and test material.
(6) Product quality audits.
(7) Review of test and inspection data.
(8) Field (on-site) performance testing.
(9) Internal testing and release.
(10) Evaluation of field stock and spare parts.
(11) Data processing inspection and test reports.

Internal Failure

(1) Scrap.
(2) Rework and repair.
(3) Troubleshooting or defect/failure analysis.
(4) Reinspect, retest.
(5) Scrap and rework—fault of vendor—downtime.
(6) Modification permits and concessions.
(7) Downgrading.

External Failure

(1) Complaints.
(2) Product or customer service.
(3) Products rejected and returned.
(4) Returned material repair.
(5) Warranty costs and costs associated with replacement.

CLASSIFICATION OF WELDED JOINTS—A KEY QUALITY ASSURANCE STEP

The International Standards Organisation Technical Committee—Welding, has proposed [13] that the factors to be considered by the designer in fixing the service requirements for each welded joint in a construction are:

(a) the degree of certainty which must be obtained that the joint will perform satisfactorily for its design life;
(b) the consequences of failure;
(c) the factors influencing the performance of the joint.

The joints can be categorised into the highest service requirements: Type 1 where joints are under 'most severe conditions and/or the failure of which would have any catastrophic consequences', down progressively to Type 4 for 'joint under non-critical conditions the failure of which would not effect the efficient performance of the construction as a whole'. This subject has been a topic for countless committees, both national and international. The International Institute of Welding, Commissions V, XI, and XV have debated the subject and their journal *Welding in the World* contains many papers and surveys on the subject. This is because, as is shown elsewhere, the costs of inspection, including NDT can outweigh the costs of welding, and since the amount of inspection is directly related to the classification of welds, the amount of inspection, the stages of inspection and the methods of inspection should be determined as a result of a detailed study of factors, generally termed 'influencing factors'. Often, as in the past, we resort to a 'code'—it is to be hoped that code writers have studied the work and papers on the subject.

The best approach to quality control comes from a sound understanding of technical factors, and nowhere is this more important than in welding. It could be argued therefore that quality control procedures should be based on such knowledge and considerable guidance given in ISO document 3088 (1975), Factors to be Considered in Specifying Requirements for Fusion Welded Joints in Steel [14].

The quality control procedures, including process controls, design checks, inspection etc. should take account of such factors as:

Materials (parent and filler)

Chemical composition, homogeneity, surface condition, thickness, etc.
Property and size of the heat affected zone.
Compatibility of weld metal.
Properties of the weld metal.

Welding processes and procedures

Must be compatible with the materials.
Differences between shop and site welding.
Profile and finish may be affected.
Heat inputs from different processes may be critical.

Stresses

Facts of fatigue must be understood.
Stress concentration factors depend on type of joint geometry, defect direction and orientation.
Residual stresses.
Fillet weld configuration.

Geometric effects

Distribution of stresses should be disturbed as little as possible by joint geometry.

Avoid severe changes in section.

Junction of thick to thin sections require special consideration.

Environment

Joints subject to corrosive or erosive environments, geometry, protective coatings, etc., require consideration.

Specifications, both for welding and quality control require that these, and all other relevant features are incorporated. Design review will ensure that they have been considered.

HUMAN FACTORS IN QUALITY ASSURANCE

Perhaps the most important factor in quality assurance is the human one, partly because its effects permeate design, manufacture, inspection and operational stages.

Further attention is given to this aspect in Chapter 6, but those characteristics that must be recognised, particularly by quality assurance managers onto whom much of the responsibility for an organisation's quality system falls, are: cutting corners, boredom, fatigue, ignorance, lack of training, ambiguous instructions, poor communications, weak supervision, failure to obey rules, concealment of mistakes, personal unsuitability for given task, carelessness and a measure of arrogance.

It will be clear that, whilst some of the factors listed above are difficult if not impossible to control, others can form part of the training programme for individuals whose work will be seriously affected by particular deficiencies, or for individuals who have displayed weaknesses. Much however can be overcome by better instructions and procedures, by management and supervisor motivation and by audits to determine adequacy of instructions and adherence to instructions.

QUALITY AUDITS AND SURVEILLANCE

No book on quality assurance would be complete without some reference to auditing since this aspect, although a key element of a

<table>
<tr><th>DATE/TIME</th><th colspan="2">TOPIC</th></tr>
<tr><td rowspan="4"></td><td colspan="2">Pre-audit conference—scope, agenda</td></tr>
<tr><td>Audit of management,
design and procurement</td><td>Audit of manufacturing,
assembly and test</td></tr>
<tr><td>Quality assurance programme
Planning and reporting
Internal and external
Audit files
Corrective action
Design control
Document control
Procurement control
Records system
Personnel training
and qualification</td><td>Manufacture planning

Receipt inspection area
Material stores area
Material handling
Process control
Assembly and test area
Inspection
Test control
Non-conformance control

Calibration
Records</td></tr>
<tr><td colspan="2">Post-audit conference—findings, agreements, commitments</td></tr>
</table>

FIG. 3. Typical agenda for a comprehensive suppliers audit (from Sayers and Macmillan, reference 15).

quality assurance system is relatively new in the field of welded construction.

A suitable definition of quality audit is—'a systematic and independent examination of the effectiveness of the quality system or of its parts.' An audit is a prime method of obtaining factual information from an unbiased assessment of objective evidence rather than subjective opinion. It can relate to products, processes (e.g. welding) or organisations. It generally includes an evaluation of the suitability of the requirements as well as compliance with them.

The effectiveness of audits depends on the co-operation of all parties concerned, and its objectives should not be confused with routine inspections and surveillance activities. Figure 3 shows a typical agenda for a comprehensive supplier audit [15].

As an example of what can be included in a quality system for welded pressure vessels the work of Commission XI of the IIW can be cited. They appointed a working group who have identified the controls and checks to be used as part of quality assurance (see Table 3).

Quality surveillance is the supervision of a contractor's quality assurance organisation and methods.

TABLE 3

CONTROLS AND CHECK POINTS IN PRESSURE VESSEL CONSTRUCTION ACCORDING TO IIW COMMISSION IX WORKING GROUP ON QUALITY ASSURANCE[a]

Controls	*Number of check points*
Examination of plans and calculations	6
Receipt and control of base materials	20
Receipt and control of consumables	20
Qualification of welding procedures	30
Qualification of welders	23
Control of work preparation before welding	4
Control during welding	15
Control after welding	20
Control of heat treatment	20
Final tests	6

[a] From a review by the International Institute of Welding Commission on Pressure Vessels (Working Group on Quality Assurance).

Below are given some of the factors to be considered by a purchaser when developing a surveillance strategy with a particular contractor. An assessment (or audit) against these features will identify weaknesses which must be the subject of special action during surveillance.

FACTORS TO BE CONSIDERED IN PREPARING A QUALITY ASSURANCE PROGRAMME

Plant Item Criticality

(See classification of welded joints).
Identification of statutory or regulatory requirements.
User experience of plant and operational problems.

Performance Capabilities of Contractor

Supplier facilities records.
Supplier performance records.
Supplier evaluation reports.

Contract Planning and Engineering

Review of contract specification requirements.
Review special customer requirements.

Examine programme control proposals.
Examine design/engineering proposals.
Examine work instructions, standards, drawings etc.
Monitor design change and concession controls.

In-house Manufacture Inspection and Test
Material verification.
Processes examination review and approval.
Personnel examination review and approval.
Monitor contractor's control of manufacture, inspection and test.
Inspections and tests during manufacture.
Final inspections and tests.

Purchasing and Sub-contracts
Review of proposed/selected sub-contractors.
Examine technical content of sub-orders.
Examine contractor's proposals for control of sub-contracts.
Monitor contractor's control of sub-contracts and press for improvement where appropriate.
Surveillance of sub-contracts.

Site Erection and Commissioning
Review of proposed site erector.
Monitor material handling, storage and control.
Monitor site manufacturing processes.
Monitor site assembly and erection. Stage final inspection and test.
Assist during commissioning and setting up of work.

In the chapters that follow, many of the above concepts and practices will be discussed in greater detail.

HOW MUCH QA?

Finally, several national codes for QA refer to 'levels' of QA. This is not to be confused with levels of 'quality'. Levels of QA can be developed:

(1) For control of the finished product (simplest).
(2) For control of processes (e.g. welding QA).

(3) For control of the organisation (including flowlines, responsibilities, documentation, non-conformance and internal audits).
(4) For control of management (internal communication, motivation, training etc.).

If there are no risks in an enterprise then no QA is required—if risks are small, level 1 might suffice. Most industries involved in welded work will want to use level 2 or 3, using control at level 4 for special situations as discussed earlier.

The chapters that follow develop the concepts that make up the content of a quality assurance programme.

REFERENCES

1. European Organisation for Quality Control, Glossary of Terms, 5th Edition EOQC Berne, Switzerland, 1981.
2. Allied Quality Assurance Publications, available from Government Offices of NATO (Defence Department).
3. Rickover, Admiral, Quality—the never ending challenge, Paper to 44th Annual Metal Congress, 1962. ASTM, Philadelphia, USA, 1963.
4. A Report of the Royal Commission on the Failure of Kings Bridge, 1962. State of Victoria, Australia.
5. Brittle Fracture of a Thick Walled Pressure Vessel, BWRA Bulletin, June 1966.
6. Report on the Brittle Fracture of a High Pressure Boiler Drum at Cockenzie Power Station, South of Scotland Electricity Board, January 1967.
7. Burgess, N. T. and Levene, L. H., The control of quality in power plant—a large user's campaign against manufacturing defects, I. Mech. E. Conference, Sussex University, Paper 12 (1969).
8. Health and Safety At Work Act (particularly Section 6), HMSO, London.
9. International Standards Organisation Technical Committee, TC-176. Refer to national standards organisation of particular country.
10. Belbin, R. M., Quality calamities and their management implications, British Institute of Management, OPN8, 1970.
11. Burgess, N. T., The development of quality assurance in major engineering projects, Paper to Institute of Quality Assurance World Conference, London, October 1980.
12. BS. 6143 Determination and Use of Quality Related Costs, British Standards Institution, London, 1981.
13. Document 3041, 1975.
14. Document 3088, 1975.
15. Sayers, A. T. and MacMillan, R. M., (1973), Auditing the quality systems of suppliers to the CEGB, *Quality Engineer,* **37** (7/8).

2

Designing Weldments to Avoid Failure

J. E. M. Jubb

Messrs Sandberg, Bedford, UK

INTRODUCTION

The design of welded connections is a more demanding task than it appears initially and the essence of the subject is that nothing is overlooked. The subject embraces the following:

(1) Use—loading and environment.
(2) Structural forms, analysis, detail.
(3) Material selection.
(4) Process(es) to be used.
(5) Inspection techniques.
(6) Defect acceptance level(s).
(7) Service control.
(8) Economic considerations.

At all stages of his design assessment the designer must be conscious that concept and reality are not necessarily the same; dimensional tolerance, affected by weld shrinkage, access for welding, variability in materials and consumables and the accuracy of defect assessment techniques are realities which the designer may ignore in the conceptual stage. This deeper and wider understanding of the subject will not be fully necessary when conventional or well-proven materials are being welded by established methods, but overall caution and insight should never be relaxed Often the experience of the fabricator overcomes the shortcomings of the design concept, especially where there is a rapport between the two parties.

Awareness of the scope of the design problem is a basic step in a fully assured system and unfortunately only a small percentage of designers understand the wider spectrum of topics listed above, and their respective implications in the manufacturing, assembly and erection stages.

It will be found in further chapters in this book that the success in performance will come from full implementation of various systems embracing conception, design in the widest sense leading to complete procedures for fabrication, inspection, installation and correct use in-service. The designer must be the person who knows the constraints on his design and is in a position to guide the fabricator and inspector on the implementation of his design concept, to see that the installation is achieved according to the plans and to advise the user on the control of functions in-service.

Designing welds against failure in the context of this chapter demands a more penetrating look at certain basic parameters, but nevertheless it would be foolish to isolate these parameters from any of the items which are discussed in other parts of this book. The welding designer should aways be concerned with detail acknowledging the effect of the detail on the whole.

The role which the designer actually takes depends on the circumstances; this could be relatively superficial in the established and proven area of industrial steel buildings, but, in contrast, the designer's role in nuclear power or nuclear submarines would be a more wide ranging and profound range of responsibilities ensuring that no significant point is overlooked from conception to manufacture, through the full service life and beyond.

It is at the design stage where roles may not be defined clearly and as a result the assurance of the total system is not established with the necessary authority. The designer needs to have the scope of his brief defined by technical management, if he is to fulfil his role efficiently and without misunderstanding by other parties in the contract. This approach is the exception rather than the rule and the neglect of this discipline can and has led to serious problems including major structural failures. The problem is particularly acute where the design is separated into several stages each involving a further sub-contractor.

A possible sequence is given:

(1) Overall layout and loading—main consultant.
(2) Analysis and choice of materials and sections—computer group.
(3) Detail design—fabricator.
(4) Manufacture—another fabricator.

Lack of a brief for the designer and strict control of the responsibilities of each sub-contractor must therefore produce areas of uncertainty and a greatly enhanced risk of an incomplete design function.

Some fabrication contracts include the detail design of the conventional joints whether welded or bolted. Joints carrying complex loads or having unconventional geometries would be the design responsibility of the original designer and not the fabricator. This division of responsibilities can lead to conflicting design and detailing philosophies.

If the design is to be defined in terms of responsibilities, this definition must occur before work starts and any checking controls must be clearly identified for the various parties involved. In practice where separation of responsibilities in the design process occur, it will be vital to establish formal links between the various parties to ensure that no decisions are taken in isolation and ignorance of other critical factors does not lead to subsequent difficulties especially in production. The versatility of welding as a joining process can encourage the designer to detail all sizes and shapes of weld and not all of them can always be produced successfully.

Without a substantial background in the technology of welding a design engineer might create problems by producing details which have limited access for high quality welding, have welding details where the heat input from the welding process is inappropriate for the properties, such as fracture toughness, which have to be achieved and which produce geometries which cannot be inspected satisfactorily.

DESIGN EVALUATION (AUDIT)—A WORKED EXAMPLE

An example has been chosen of a heavily stiffened plate, see Fig. 1, which presents a variety of fabrication problems. Orthogonal stiffening at close pitches in both directions presents a challenge for the full penetration butt welding of the through stiffeners and an even greater challenge for the short lengths at right angles which according to the drawing are to be welded all round. Lamellar tearing of the plate due to weld shrinkage of the cruciform joints is a possibility, where consideration would need to be given to the use of steel with enhanced through thickness ductility.

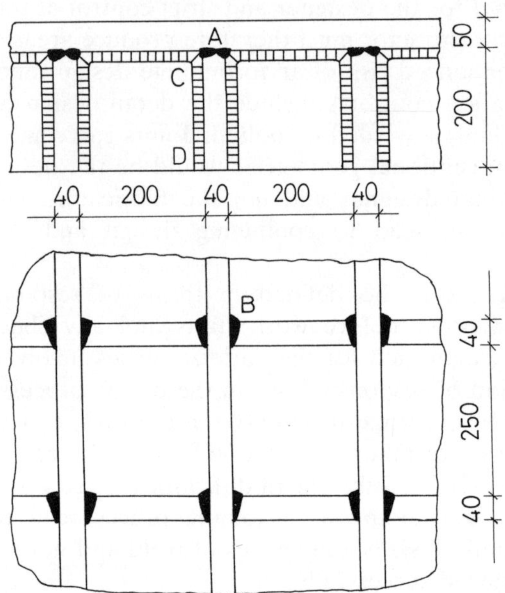

FIG. 1. Butt welded stiffened plating.

The fabrication is criticised in the following section and an alternative solution is presented.

Criticisms

Full Penetration Butt Welds at A and B

Are full penetration welds essential? Full penetration welds give maximum shrinkage and restraint during welding. Access between stiffeners may be limited for welding and sufficient access will depend on the process being used (manual metal arc, MIG CO_2, flux cored MIG, submerged arc, electroslag) and the position in which it is to be welded. More access could be necessary for back chipping or gouging out the root of the weld than for welding depending on the spacing and depth of the stiffeners.

Welding Round Corners, C, Over Existing Welds

In principle there is no obvious problem with this type of detail, but in practice the following reservations need consideration. When welding round a corner or into a corner and out of it there is a significant risk that a defect due to slag entrapment or shrinkage may occur at the corner. It is very difficult or impossible to non-destructively test a weld corner such as C. The weld in the corner cannot be guaranteed for quality. The thermal stresses and strains of the corner weld may aggravate any existing shortcomings in the adjacent butt weld on metal or heat affected zones.

End of Butt Weld at D

If run-off plates were introduced at D, they would be removed by flame cutting and would require many hours of grinding back to produce an acceptable end profile. This arises due to the similar depths of the stiffening members.

Single Bevel Butt Welds at B

As drawn the butt welds at B are made from one side and would require a backing strip under the weld to guarantee full penetration, when produced under normal commercial conditions in a fabricating shop.

REVISED DETAIL FOR ORTHOGONALLY STIFFENED PANEL

The revised detail shown in Fig. 2 is an alternative form of construction which meets some of the objections raised in the discussion of the fully butt welded original scheme, but there is no claim that this is an optimum solution which meets design, production and inspection requirements. As with most complex forms of construction there has to be some compromise between conflicting requirements, but the options need to remain open for the fabricator to select an economic solution within the constraints imposed by design/service requirements. The main change is from butt welded to fillet welded connections.

The removal of the internal corners of the intercostal stiffening plates must be sufficiently generous to allow welding through the cope hole (mousehole or snipe) if this is deemed to be necessary. For thinner plate a 25 mm radius is a sensible minimum, but where the weld quality of the end of the welds finishing at the edge of the cope hole is critical a more generous size of cope hole would be necessary

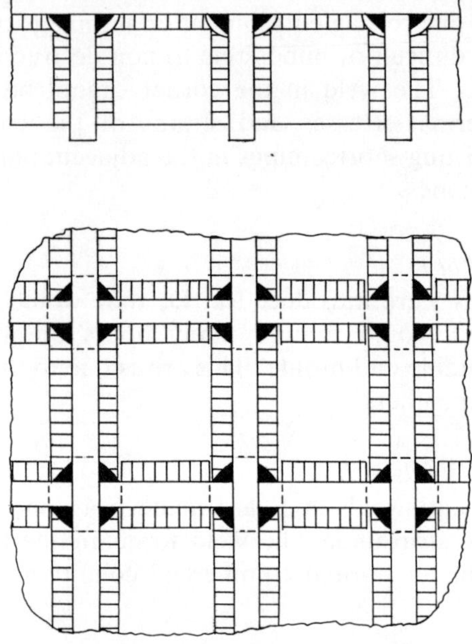

FIG. 2. Fillet welded stiffened plating.

for satisfactory welding and inspection. This matter should be discussed with the fabricator. The thicker the stiffener the more generous will be the requirement for the cope hole, if quality of the ends of the welds at the cope hole is to be sustained. It should be noted that the cope hole should preferably have a smooth surface with the rags removed and this will be more critical if the joint under discussion is subject to fatigue loading. It is good practice to dress a detail of this kind in order to obtain the most effective results from the protective treatment. Access for protective treatment is difficult in corner positions between stiffeners at right angles and uneven flame cutting and rags may not be prepared thoroughly leaving the risk of impurities under the protective treatment and the possibility of accelerated corrosion in-service.

Full penetration butt welds have been replaced by fillet welds reducing the through thickness strains. Reducing the depth of the shorter stiffeners allows the welder to carry the weld round the end of the stiffeners and no dressing is necessary at that position unless the weld profile is very poor.

Having introduced a cope hole there are a large number of relatively short runs of fillet weld terminating at the end of the cope hole. This could prove an untidy detail unless the welder is very careful and, if it was lumpy at the end of the weld, it could prove difficult to inspect. In fact cope holes become more problematical the thicker the plate.

It is seen from the above discussion of the two methods of construction that neither is ideal and there is room for further simplification and improvement in detail design. Further options could be explored; depending on the economics a casting has much to recommend it with continuity and simplicity. Returning to welded solutions it might be possible to increase the pitch of the stiffeners to improve the access for welding. It would be advantageous to eliminate the short transverse stiffeners and this option should be explored in great detail. Welding design requires evolution and compromise; the result must be a solution for which the necessary quality can be achieved.

DESIGN AGAINST FAILURE

Failure may bring to mind shattered pieces of metal lying on the ground. Pictures of shattered aircraft which have crashed are familiar examples in the newspapers. In the majority of failures in welded construction the results are far less dramatic and in these terms a definition of failure is appropriate. Failure occurs when the fabrication no longer meets the requirements of the design specifications and the component needs repair, ongoing monitoring in-service or replacement. The failure could be overall but it is usually local; there might be a massive brittle fracture running through a vessel or pipeline at several hundred metres per second or in contrast dimensional tolerances might be lost in a precision machined component due to local yielding at a welded joint, where the applied load has caused a partial mechanical stress relief of the welding residual stresses in an as-welded component and some permanent set.

The basic ductile strength of a steel fabrication is rarely a problem, unless the welding is not in accordance with the drawing with undersize or defective welds, because the majority of welded steel fabrications have the weld metal overmatching the ductile strength of the adjacent parent plate. Hence in less critical steel fabrications where inspection may be at a relatively low level or non-existent, the risk of the combination of poor quality welding and the circumstances to bring about the designer's maximum loading conditions is relatively low. In other materials or much higher strength steels where the weld metal or heat affected zones may not match the parent plate strength more care will be required. Also, highly stressed connections such as welded joints in tubular steel trusses with local magnification of stresses due to secondary bending and variable stiffness in the connections due to rapid changes of shape, require more detailed attention.

A further factor in the very low incidence of welding failure through ductile rupture is the traditional engineering education which, in design courses, concentrates on strength and stiffness. Usually course work is based on strength under static loading conditions with a factor on yield stress or ultimate strength related to ductile fracture. Stiffness is checked to avoid unacceptable deflections and possible local damage to components attached to the fabrications. Where tight dimensional tolerances are required, further consideration will have to be given to the manufacturing process. Shrinkage stresses associated with the steep temperature gradients of the weld thermal cycle and uneven cooling are of yield point magnitude in tension and any subsequent tensile strain will give rise to local plastic flow and a small amount of permanent set. If the permanent set is unacceptable within the finished tolerances, the designer will need to call for post-weld heat treatment or, rarely, preloading for a fabrication always loaded in the same direction to reduce the level of residual stresses to a level where there will be no possible dimensional change of the fabrication in service.

A close study of the welding literature reveals that brittle fracture and fatigue are more dominant issues in design and failure than ductile strength and the failures reflect the neglect of both subjects in many engineering courses. Now the demands of the offshore industry with structures in hostile environments producing low temperatures and severe and repeated wave loadings have given further emphasis to the need for defining these modes more accurately in the design process and despite considerable research effort in the past three decades further funds are being spent to provide even more background to the multiple variables involved.

Brittle Fracture

Brittle fracture in welded steel structures can take place at stresses considerably below the yield or proof stress and at room temperature.

Brittle fracture is generally more critical for:

increased thicknesses;
higher strain rates;
lower temperatures;
sharper and larger notches;
lower fracture toughness in parent plate;
weld metal;
heat affected zones;
higher levels of residual stress.

The number of variables is a challenge for the designer and in particular an understanding of the role of fracture toughness in brittle fracture is essential to a correct interpretation of the data. Codes and standards cover basic requirements for materials and matching properties are sought for the weld metal and heat affected zones. Typically Charpy V-notch values have to be achieved at relevant 'design' temperatures, but, where better definition is necessary, welding procedure tests may be assessed by the crack tip opening displacement test although the cost is much higher. Any departure from the norm with the use of new materials and new welding processes/procedures will require a reassessment of fracture toughness and the possible need for further tests to define the new production conditions more closely. Engineers wanting to undertake pioneering work in fabrication need to study welding technology in depth or take advice. Resistance to brittle fracture is vital and a feel for the nature and scope of the problem at the conceptual stage is essential.

In fabrications where structural safety is very critical the designer must make searching enquiries on the capacity of non-destructive testing to define the size of defects which may occur in the various parts of the structure. It has to be recognised that non-destructive testing has certain limitations depending upon the technique and the operator, and that definition of defect size is subject to certain physical limitations as well as organisational ones. This knowledge is an essential part of a strategy to ensure adequate resistance to brittle fracture.

Fatigue

Fatigue cracking, in stark contrast to brittle fracture, is a slow propagating process. Whilst brittle fracture has been eliminated to a large extent from welded construction by a suitable choice of materials and consumables, fatigue cracking remains a continuing source of trouble in-service by its very nature, where repeated loading occurs. Brittle fracture usually describes a single event which takes place very quickly due to a combination of circumstances which cause a lack of equilibrium. Fatigue cracking is an accumulation of events running into hundreds of thousands or millions taking place in most cases over several weeks, months or years; the sequence and scale of the events may be important together with variations in the environment. The designer is expected to anticipate all the possible variables and integrate them in a design method. There must be a high degree of approximation in the solution of fatigue life taking into account the random nature of the events in most fabrications in-service.

The criteria for fatigue failure could be many. Some examples are quoted:

(1) A fatigue crack not penetrating the thickness of the material, but sufficiently large to initiate a fast brittle fracture.
(2) A fatigue crack through the wall of a vessel or pipe allowing leakage of the contents.
(3) The item is used until it breaks due to fatigue cracking and is replaced. If small fillet welds are used, the fatigue crack could propagate from the root of the weld through its throat and not be found until the crack appeared on the weld face causing failure. For larger lower stressed fillet welds cracking usually starts at one of the weld toes and can be found by a method of crack detection.

The most common method of assessing potential fatigue failure is the linear cumulative damage rule or Miner's rule. The rule assumes that every cycle of loading makes a fractional contribution to the growth of the fatigue crack, although more recent studies indicate that there is a cut-off for small stress range cycles. For practical purposes a designer can simplify the applied loading to obtain an approximate solution for the fatigue strength; this can be achieved by breaking down the variable loading into a few groups or stress ranges and estimating the number of events in each group. In the majority of designs, fatigue will not be the mode of failure and the approximate analysis will prove sufficient to eliminate it from further consideration but should fatigue be found to be a potential problem more searching checks will be necessary. Modification of the actual weld detail to a category which is less sensitive may prove sufficient or moving the critical weld detail to a region of lower stress may enhance the fatigue performance; however where one tube is being welded at an angle to another tube, whether circular or rectangular, the options are strictly limited and the stress/strain concentrations must remain quite high. Tests show that welded tubular connections have a very low fatigue strength and designers need to keep this relative weakness in mind. Attention to local detail may enhance the fatigue performance, but it cannot alter the high stresses and strains arising from an abrupt change of stiffness and uneven distribution of load transfer in a welded tubular connection.

Quality control in construction is an important related variable. Fatigue design is based on full integrity of the material and the welds joining it. In highly stressed structures it will be necessary to ensure that shortcomings of the production processes do not lower the fatigue strength. Arc strikes made carelessly on the surface of the material away from the weld preparation may give rise to small crater cracks which could act as fatigue crack initiators. Insufficient control of preheat and consumables may lead to moisture contamination of the weld and the associated evolution of hydrogen could cause cracking in the heat affected zone and a possible reduction in the fatigue life. The designer may feel that issues arising in fabrication are outside the scope of his responsibilites; this is a mistaken idea. Unless the inspector has clear backing from the technical specification of the quality which is required, the fabricator may resist remedial works which are considered to be necessary to achieve a satisfactory fatigue performance. The fabricator will be looking for extra cash and the designer may be forced to compromise with unfortunate long term consequences.

Where fatigue performance is known to be critical, much effort should be devoted to the specification to ensure that the final product will meet the concept of the designer in every respect. In very critical circumstances a pre-tender meeting should be held with the various fabricators to inform them verbally of the requirements in order that they can allow for the extra costs involved in supervision and remedial work. A tender will be presented on various assumptions and the tenderer will be influenced largely by his previous experience. If the purchaser is seeking a solution in terms of quality which is foreign to the track record of the particular fabricator, the purchaser must decide that this is the case and inform the fabricator accordingly. If this situation is not recognised and discussed pre-tender, there is the strong possibility that the fabricator will proceed along his normal path with some minor adjustment and the purchaser will recognise eventually that the solution, the fabrication which has been produced, does not meet his exacting requirements. This is a well established formula for aggravation and recrimination.

There is a further point which should be made to the designer. Fatigue performance is very sensitive to the stress level and a small reduction in stress level may lead to a substantial increase in cycles to failure. This places more emphasis on the accuracy of the loading and particularly in a complex joint on the accuracy of the analysis. It has been emphasised that there should be a reasonable balance of effort between a better understanding of the loading and the analysis. No amount of refinement of the analysis of the stress distribution will compensate for a lack of understanding of the range and frequency of the applied loading.

An appreciation of the relative merits of various types of welded joints in a fatigue environment can be found in various codes and standards.

Other Failure Modes

The performance of welded joints has been reviewed briefly in terms of ductile strength and stiffness, brittle fracture and fatigue cracking, the most common modes of failure. The designer cannot be content that he has now covered all the modes of failure; there are other aspects of behaviour which merit attention and need to be reviewed by the designer. A continuation of the list of modes of failure is:

(1) Instability affected by reduction of the buckling strength of welded plate elements.
(2) Creep behaviour at high temperatures.
(3) Environmental factors,
   (i) corrosion
   (ii) stress corrosion
   (iii) wear
   (iv) erosion
   (v) wind excited vibration
   (vi) wave excited vibration
   (vii) earthquake accelerations
   (viii) noise excited vibration.
(4) Interaction of various modes of failure mentioned previously.

It is not claimed that the above list is exhaustive but it gives an indication of the range of problems which need to be eliminated at the conceptual stage of the design.

Item (1) on the list covering the reduction of the buckling strength of welded plate elements is usually a secondary issue. Taking a simple box column as an example, made from four steel plates welded at the corners, there is residual longitudinal tension at and close to the welds at the corners of the box. This arises due to the welds wanting to shrink on cooling and the adjacent plate in the faces of the box columns resisting the shrinkage. To achieve equilibrium of a locked-in system of welding residual stresses the plates between the welds at the corners carry longitudinal compressive stresses. In practice the tensile shrinkage stresses at the corners are at yield level, but the compressive stresses providing the equilibrium are spread over a much larger area and are at a much lower level. It is the pre-stressing of the majority of the plate element in compression which reduces its capacity to carry as much compressive stress as the designer might expect. This type of behaviour could be more acute in an I-beam welded from three plates with longitudinal tension associated with the weld at the web to flange connection and with the possibility of longitudinal compression at the free edge of the flange, the more susceptible element to buckling.

Creep behaviour of welded joints, Item (2), is a specialist subject and needs to be discussed with parties who have experience with this subject and who have access to relevant experimental data.

Environmental factors are numerous but in a particular design it is possible that few or none may be relevant to the mode of failure. Probably the two environmental categories which merit the most attention are the various kinds of corrosion and the dynamic behaviour brought about for a variety of reasons and which may prove unacceptable.

BIBLIOGRAPHY

Journals for general reading

Welding in the World, International Institute of Welding, The Welding Institute, London.

American Welding Journal, American Welding Society, New York.

Metal Construction, The Welding Institute, London.

Brittle Fracture

BS DD55 : 1978 Fixed Offshore Structures.

BS 4360 : 1979 Specification for Weldable Structural Steels.

BS 5762 : 1979 Methods for Crack Opening Displacement (COD) Testing.

BS 5500 : 1982 Unfired Fusion Welded Pressure Vessels.

Fatigue
BS DD55:1978 Fixed Offshore Structures.
Gurney, T. R. *Fatigue of Welded Structures,* 2nd edition, Cambridge University Press, Cambridge, 1979.
BS 5400:Part 1:1980 Steel Bridges—Code of Practice for Fatigue.
ANSI/AWS D1.1–82 Structural Welding Code—Steel.

Ductile Strength
Blodgett, O. *Design of Welded Structure,* Lincoln Arc Welding Company.

3

The Control of Quality During Shop Operations

A. F. GIFFORD
Northern Engineering Industries (Mechanical Engineering Ltd), Derby, UK

INTRODUCTION

Over the past few decades the demand for reliability of engineering systems has increased significantly. This has been a direct result of the ever-increasing cost of 'down time' of major capital plant, where loss of revenue can often exceed £30k per day. In the recent past there has also been a number of major failures of structural components and pressure parts which have become headline news, sometimes resulting in loss of life. Such failures include Flixborough Chemical Plant, Milford Haven Bridge, the Alexander Kieland offshore accommodation rig; their cost runs into millions of pounds. Fortunately, the number of such multi-million pound disasters is limited, but every year equally large sums of money are lost in fabrication workshops due to welding problems. Welds or parent materials crack due to a variety of reasons; porosity and slag makes seams unacceptable; distortion renders components unusable without major rework; unsuitable welding consumables are used or specified properties are not achieved. Problems such as these are encountered every day in industry and result in loss of profit to the fabricator, loss of reputation and often associated delay to major projects. Many of these problems can be avoided if the fabricator follows a systematic course of quality assurance at all stages of his operation.

Primarily companies are in business to satisfy their customers, to make profits in order to satisfy their shareholders and to ensure the continuity of the company. It is however essential in achieving a profit that the correct balance is obtained between the cost of actually manufacturing the products, delivery to a predetermined programme, and ensuring that the quality is as specified. The important feature is to control the balance between these key functions.

This chapter will consider the inter-relationship which must exist between the various parties associated with fabrication. This may be very complex in the case of major projects involving the ultimate client, consultants, insurers, and major sub-contractors, as well as the fabricator himself and his sub-contractors. A simplified route illustrating the basic relationship from client to operation is shown in Fig. 1.

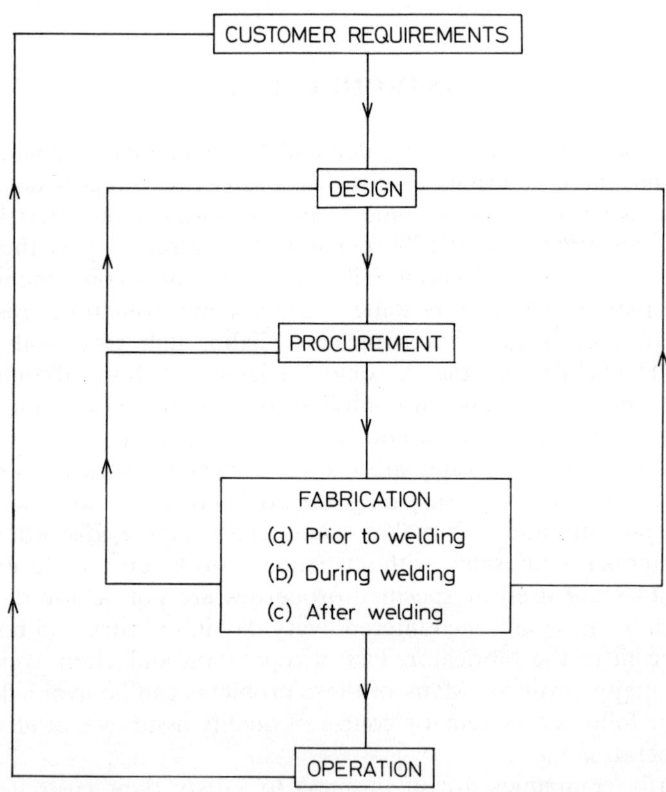

FIG. 1. A simplified outline of the QA route to be followed in welded fabrications.

DESIGN CONSIDERATIONS

The importance of design in respect to production costs is paramount. The designer must specify clearly and precisely what shape he is seeking to be fabricated, what materials are to be used, and what levels of dimensional control and weld quality he is seeking. Far too frequently the designer backs off from defining the limits applicable to fabrication, preferring to leave this to the shop floor, since he often does not understand *what* can be achieved economically.

To enable a weldment to have its required reliability throughout its working life it must incorporate, as a minimum, the correct level of quality—quality being defined in this instance as 'fit for the intended purpose'. There is no bonus for over-design or excessive quality being provided, and adequacy should be seen as the minimum requirement to be achieved, even though in certain cases that adequacy may constitute near perfection.

The role of the designer in respect to welded components can be defined as:

(a) Determine the functional and material requirements, observing any requirements of relevant codes.
(b) Translate these into working drawings and specifications which incorporate value engineering, production and quality engineering, reliability and research and development, as appropriate.
(c) Establish tolerances and acceptance criteria.
(d) Design the joints such that access for welding and testing is available.
(e) Operate within economic and programme constraints.

The part played by the designer in the success of a fabrication is most important and, to ensure his conformity, a 'design review' is desirable in many cases. Such reviews should be held at three stages:

(a) product definition,
(b) basic conceptual design,
(c) detailed design.

The meetings should be attended by representatives, as appropriate, of sales, quality, production, welding, procurement and other involved departments. It is also desirable that a non-involved designer is present to ensure a detached viewpoint. The objective of such design reviews should be to:

(a) ensure specified customer requirements are met;
(b) relate to previous experience;
(c) review welding, inspection and testing requirements;
(d) design details suitable for fabrication;
(e) identify requirements for new processes;
(f) record the decisions and reasons.

Many will consider that such meetings will delay the release of information to manufacturing. There is little doubt that such actions will reduce overall costs and manufacturing time, and increase the conformity to design. Costs are relatively low at the design stage and become progressively higher as the contract proceeds.

In considering the economic and programme aspects of a fabrication, it is necessary to remember that there is no clearly defined relationship to be observed. A decision taken to reduce the manufacturing time span may involve higher costs depending on a variety of factors including numbers to be made, availability of plant and materials, supply of labour, etc. Some of these issues are directly influenced by the designer who must therefore take every care to ensure he makes the best decision in respect to function, time scale and economics of the fabrication.

There must therefore be close co-operation between the designer and the welding engineer to ensure that the foregoing requirements are met. The welding engineer should keep the designer informed of developments in welding equipment and techniques which may affect the industry and he himself must be able to take the full advantages of new materials which may improve the effectiveness or economics of a design. He must determine the production route to be followed for the various welds and must reach a compromise between welding and preparation costs (i.e. large tolerances on preparation may reduce the cost of that activity but significantly enhance welding cost).

It is important that the simplest and most ductile base material is chosen which will fulfil the design requirements [1, 2]. The choice of over sophisticated materials may initially reduce the estimated cost, but production problems and maintenance of very critical fabrication requirements may well override. The choice, for example, of carbon steel will invariably prove satisfactory since it has good weldability, often together with adequate mechanical properties. The use of alloy steels may be essential in some applications due, for example, to creep

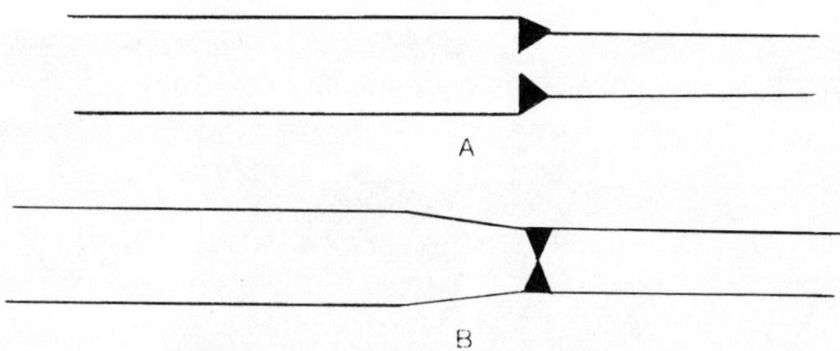

FIG. 2. Examples of section changes in high yield steel. Type A is not acceptable due to lack of penetration of weld and high stress concentration at the toe of the weld. A satisfactory design is shown as Type B.

or stress corrosion environments, but the use of such steels increases the risk of weld or heat affected zone cracking, and of the wrong material being used either in the product or weld metal. The weldability of the materials of construction is of prime importance and should be carefully assessed before any material not previously welded by the manufacturer is introduced into a design. The designer must specify the minimum strength of weld metal required for each joint, since only he knows the load it has to carry. In general he should choose the lowest strength the design will allow, but the higher the strength of the material and the greater the applied loads, the more the design must be refined. Abrupt changes in section, which may be acceptable in low strength fabrications, cannot be tolerated in high yield steels (550–700 N mm^2) (Fig. 2) and butt joints carrying tensile or pulsating stresses must be full penetration with the weld reinforcement blending smoothly to avoid notches [3].

Access for Welding and Testing

Insufficient attention is often given at the design stage to the question of access for welding. Poor access (Fig. 3) will lead to defective welding and, in such cases, the welding process or the welder is often blamed for the deficiency. In many cases models or, better still, full scale replicas (Fig. 4) are essential to ensure access exists [4]. Access is not well defined but will include [5]:

(a) Operator—this means there must be physical space into which the operator can fit in order to make the weld.

(b) Visual—for all welds made by a manual process, the operator

FIG. 3. Inadequate welding access in a large fabricated bridge structure arising from lack of interfacing between the draughtsmen responsible for different sections.

must be able to see all of the weld. In the case of automated welds this is not so essential.

(c) Technique—to ensure that the chosen technique can be correctly performed, e.g. two-handed processes such as 'TIG + filler' have different space requirements to manual electrode welds.

(d) Equipment—in some cases, e.g. MIG, the size of the equipment may limit its application in confined spaces.

For most manual purposes the operator's head and screen can be assumed to be 300 mm diameter with the eyes at least 100 mm from the top of such a sphere. Welding should not be, in general, more than about 600 mm from the eyes of the welder, whose body should be assumed to be 450 mm diameter.

When the weld is completed, access for testing is required and for dye penetrant and magnetic particle testing this does not usually present problems. X-ray testing may be impossible due to the size of the equipment and stand-off required, whilst gamma radiography can present health hazards in confined spaces. Unless joints have been designed with ultrasonic testing in view [6] this can be impractical, or take more time, not only due to space considerations but also to insufficient planar surface from which to scan the joint with probes, or the weld preparation angle itself (Fig. 5).

Care and time spent at the design stage will significantly influence the quality of fabrication subsequently obtained. Errors made at this stage may be repeated many times before they are detected, and subsequent inspection stages may well be incapable of detecting some design errors.

FIG. 4. Full scale model of lower end of ammonia converter. This rig was built to provide access for welding, cutting, testing and heat treatment to facilitate replacement of inlet forging.

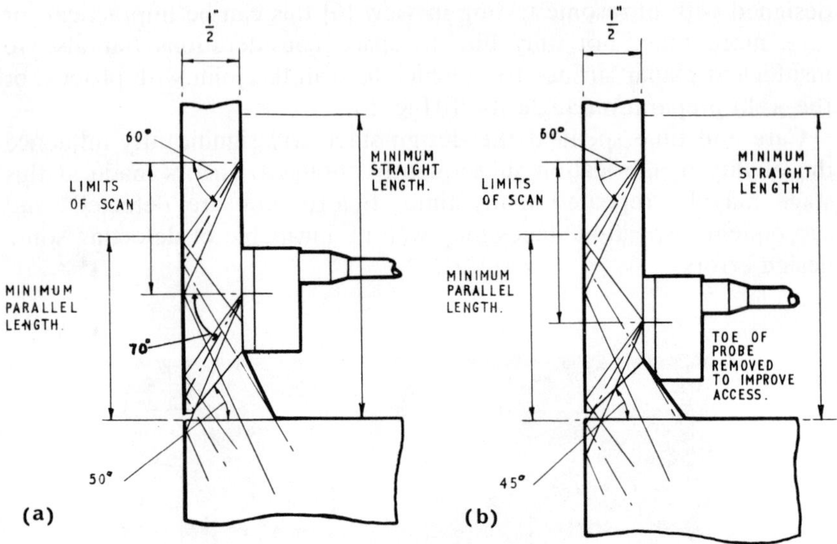

FIG. 5. An example showing how the choice of weld preparation angle can increase ultrasonic testing time by about one-third. (a) Two scans required, one at 70° for root and one at 60° for remainder. (b) One scan required at 60°. Time to test reduced by about one-third.

CONTRACTING AND ESTIMATING

The most important aspect of any contract is to determine exactly what the fabricator is undertaking to manufacture. If this is not clearly defined and understood by all parties no amount of effort in the workshops can make the activity a success. The requirement for reliability, previously referred to, together with the need to operate in a profitable manner, make the specification of the work to be undertaken of paramount importance. Words such as 'water-tight welds', 'good fusion', 'good clean metal', 'free from slag', 'free from undercut', or one-sided clauses such as 'shall satisfy the engineer' or, again, 'all welds shall be radiographed' cannot form the basis of an effective contract. The standards defined by such terminology are impossible to cost and the fabricator cannot hope to operate efficiently if they remain in his client's order.

The estimating department of the fabricator's works, in the quest for assured quality and profitability, must therefore closely review the enquiry document, isolate clauses such as those illustrated above, and tell the prospective purchaser what he will be offered, in clear, unambiguous wording, and then price his operation to suit. Unless his offer is made in such terms the purchaser can take a position whereby the work produced is never acceptable to him because he has been clever enough not to specify his requirement in detail! The route followed in handling an enquiry by a large fabricator is shown in Fig. 6.

In supporting the estimating department, the welding engineer must provide data on the processes to be used, the joint forms to be adopted, and the tolerances that must be achieved. He must also identify what

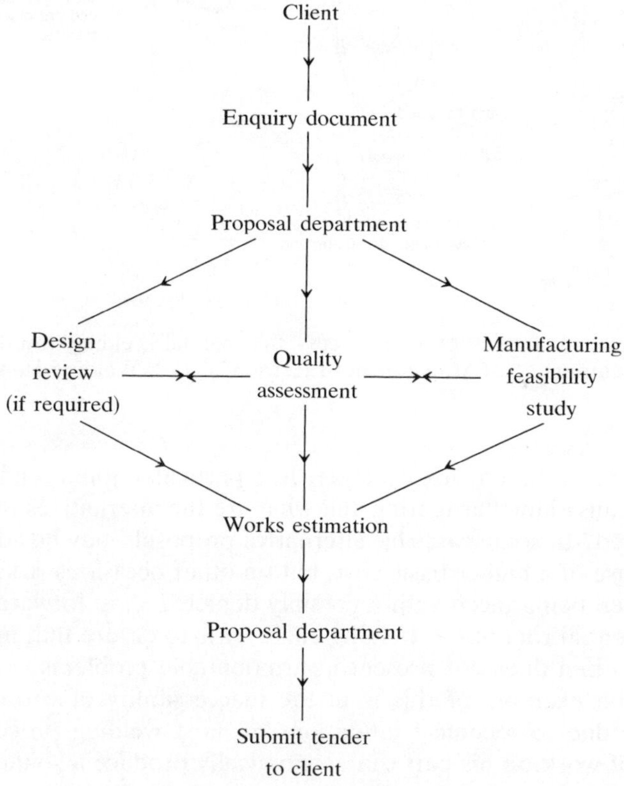

FIG. 6. The sequence of operations in producing an estimate for fabrications adopted by a large manufacturer. Welding assessment is included in the manufacturing feasibility study.

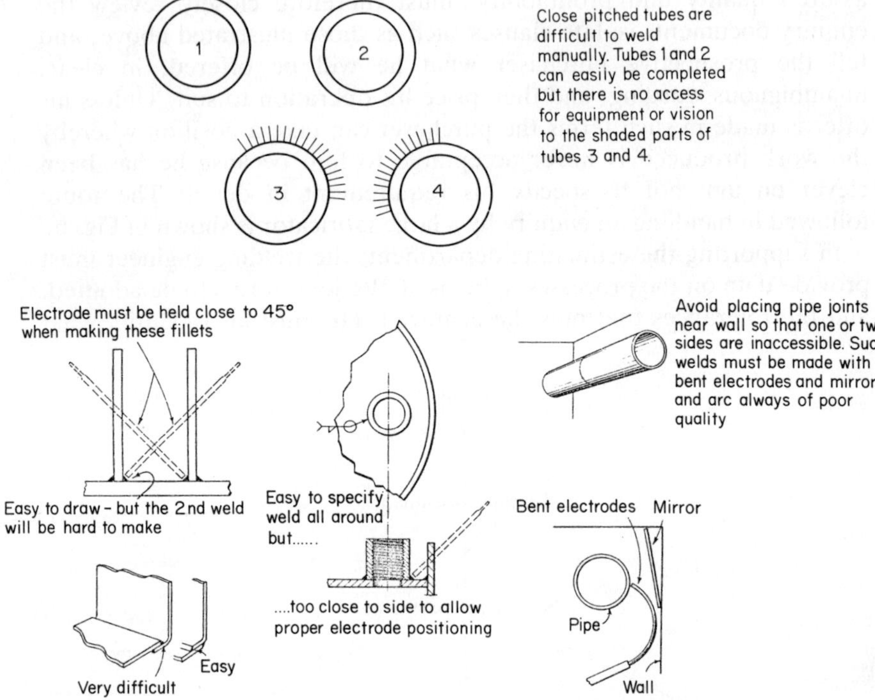

FIG. 7. Some examples of poor access for manual welding resulting from design requirements. (After Lincoln Electric Co. Ltd, Welding Handbook.)

risks are being taken; for example, is a particular joint configuration likely to cause lamellar tearing and what are the alternatives that could be followed? In some cases his alternative proposals may be adopted at the expense of a higher basic cost, but on other occasions risks may be taken when being faced with a possibly depleted shop forward load.

An essential role of the welding engineer is to ensure that the design offered to him does not present insurmountable problems to execute. A common example of this is in the inaccessibility of certain welds whereby, due to technical limitations of most welding processes, no amount of work on his part can economically produce a sound weld in the nominated location (see Fig. 7).

In arriving at the welding costs to build into an overall estimate, most organisations use their own standard data as a basis. This is then modified, as advised by the welding engineer, depending on the complexity of the component to be fabricated, and any difficulties, e.g. distortion, which are envisaged. Data on welding deposition rates for various processes is available from a number of sources [7, 8].

It is a general rule that the better the design and estimate, the greater is the chance of a successful fabrication being made.

PROCUREMENT AND MATERIAL CONTROL

Obtaining the materials and welding consumables specified by the designer or engineer at the most advantageous price and at the correct time is the role of the procurement department. They should receive clear and unambiguous requisitions defining exactly what is wanted, and must not deviate from that requirement unless a concession has been approved by the originator of the requisition. The precise definition of what is required is not always easy: British Standards for example use many clauses where 'the purchaser shall nominate . . .'. Unless these options are defined the supplier is at liberty to provide what he will—and it may not be what the purchaser requires. Problems were encountered some years ago when submerged arc welds were found to be low in tensile properties. Subsequent investigations showed that whilst there was a maximum carbon content of 0·12 % in the specification for the *wire,* there was no minimum requirement, and when 0·03 % carbon was supplied the required strength just could not be achieved. To assist in the options possible in the purchase of plate the British Steel Corporation has issued a check list (Table 1) defining everything required for complete specification of plates.

Many companies operate an 'approved vendor' register and all suppliers are assessed [9, 10] prior to inclusion. Depending on the rating of the supplier, inspection of materials may be either at the supplier's works or on receipt by the manufacturer. It is normal however, as a minimum, to check base materials on receipt for dimensional conformity, surface quality, pitting and flatness. It is also essential that any test certificates are verified for correctness and completeness and also compared to the material identification.

Control of Consumables

Welding consumables are normally purchased to national standards but in critical applications, e.g. high strength low allow composition,

TABLE 1
PLATE ENQUIRY CHECK LIST[a] (Courtesy of British Steel Corporation Plates)

	Check	Notes
(1)	Project/application	Useful in confirming the specification or suggesting an alternative, particularly important when fabricators intend to carry out cold spinning or hot forming processes
(2)	Steel specification	Including grades within a specification
(3)	Modification to mechanical properties	(a) Yield stress (b) Tensile strength (c) Elongation (d) High temperature properties (e) Charpy V-notch (or other); impact properties—indicate longitudinal or transverse (for thin plate, confirm that subsidiary test pieces and their associated impact strengths are acceptable) (f) Hardness (g) Others
(4)	Chemical composition	Indicate product or ladle analysis and any additional elements not covered by the specification. Modifications to existing specification (e.g. carbon equivalent maximum)
(5)	Heat treatment	
(6)	Plate dimensions/tonneage Plate size Plate tonneage/numbers of plates Tolerances	Gauge, flatness, length, width—confirm whether the tolerances associated with the specification stated in (2) apply; if not, give full details of tolerances
(7)	Special surface standards and treatment	(a) Surface standard? (b) Primer to be used?
(8)	Testing and inspection heat treatment of coupons frequency of testing special mechanica? special mechanical tests ultrasonic standards required private inspection	For stress relieving heat treatments, state whether these are to be shop or site and indicate soaking times and temperatures. State whether mechanical properties are to be met in the stress relieved condition. Batch test, individual test, 'as rolled' or in the 'as-heat-treated' condition? (a) Which test? (b) Number of tests per plate? (c) Test positions? (d) Specify acceptable results Levels and stages of inspection—witness tests only, or witness tests and surface inspection? State whether inspection will be by one or more inspection authorities. Please indicate which authorities. Is a quality assurance agreement to be operated? Is a quality plan required?

TABLE 1—*contd.*

	Check	*Notes*
(9)	Marking and documentation	
	marking	Any special stamping, stencilling or colour or other coding
	documentation	Any special requirements
(10)	Order/delivery	
	closing date for receipt of quotation	
	date order will be placed	
	required delivery date	
	priority sequence of deliveries	

[a] These principles are applicable for many other purchase situations.

creep conditions, etc., pads of metal are often deposited and checked on a direct reading spectrograph. Welding wires can readily be fused under argon to produce a small ingot suitable for processing on these instruments [11].

Many problems in fabricated plant arise from the use of incorrect base material or electrodes [12]. Such problems can be just inconvenient and costly or sometimes dangerous to life and limb.

The storage and control of welding consumables requires special attention. The problems associated with the presence of hydrogen in welds and its almost inevitable cracking tendencies are well documented elsewhere [13].

The avoidance of such problems calls for close control and discipline in respect of both storage and use of most welding consumables [11]. Fabricators use electrodes and other consumables from various suppliers, of differing composition and requiring different treatments prior to use. These may range from very precise baking at 450°C followed by storage at about 150°C, to others requiring little or no protection other than that offered by the maker's packaging.

On receipt in a factory all previously nominated consumables (possibly excepting simple rutile carbon steel electrodes) are placed in a bonded area until sampled and tested. Once accepted they can then be stored, preferably in heated insulated rooms with a maximum relative himidity of 55 % at a temperature of 16–25°C. Shelf life for coated electrodes is not well defined, but Table 2 forms a good basis for guidance.

There are no hard and fast rules regarding drying and baking

TABLE 2

SHELF LIFE FOR COATED ELECTRODES

Type of package	*Heated room*	*Non-heated room*
Unsealed and damaged	—	Not stored
Sealed[a]	1 year maximum	Should preferably not be stored but in any case not to exceed 6 months
Hermetically sealed[b]	3 years maximum	1 year maximum

[a] Sealed describes containers which, after opening, are used again after closing by staples, adhesive tape, etc., or cardboard type boxes; [b] hermetically sealed containers are cardboard boxes coated with plastic film, soldered tins, sealed plastic boxes, etc.

practices, and the manufacturer's recommendations should be observed. In many cases electrodes are stored in holding ovens at about 130–50°C immediately prior to use and issued into heated quivers, labelled to indicate the batch number concerned, in quantities not exceeding 4 h normal supply.

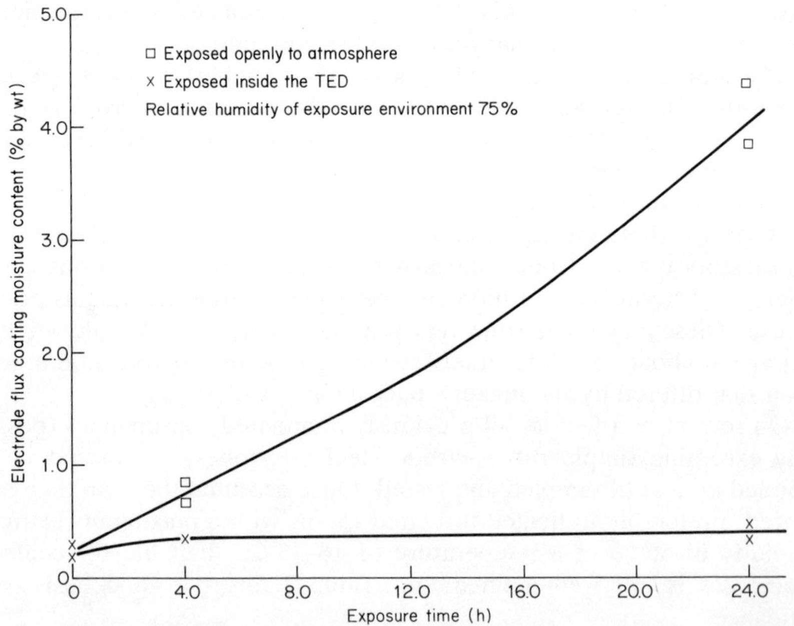

FIG. 8. Comparison of moisture pick-up of E7016 consumables inside the TED (thermal electrode dispenser) and exposed to the atmosphere.

NEI International Combustion Ltd

Date 17.1.81

Clock Number: 210	Foreman's Signature: K. Jones.
Job Number: B0478.	Procedure Number: SPECIAL 640

Electrode Type: BABCOCK & WILCOX AZ.

Rod Size	Batch Number	
14		
12	5118	
10	~~5475~~	6273
8	5381	
6		
4		

OM 50476/2

FIG. 9. Control card issued to a welder to authorise withdrawal of electrodes from the welding stores. The batch number is completed by the storeman and changed as the batch changes.

A recently marketed electrode container does not involve any electrical supply at the point of use. Electrodes are placed in a plastic sealed container, incorporating an opening flap and ejector mechanism, directly from the holding oven. The object of the system is to prevent moisture coming into contact with the already dried electrode. Independent trials have shown (Fig. 8) that moisture ingress at 75 % relative humidity is not significant and the device is considered to be viable for at least 8 h's use [13].

The storekeeper should maintain records of all issues of welding consumables (Fig. 9) and unused electrodes must be returned to store at the end of each work period, for drying. It is a dangerous and often costly practice to permit electrodes and other consumables to be littered around the workshop since they are then available to be used on critical applications in an uncontrolled manner, i.e. use of damp low hydrogen electrodes causing cracking [14] or use of the wrong consumables in a particular application.

QUALITY PLANS

The concept of planning to ensure that quality is achieved is not new. Various methods have been used in the past including inspection schedules, inspection route cards and quality control sheets. These tended to indicate what inspection and testing should be carried out on a given fabrication but, in general, they all failed to integrate manufacturing, internal inspection and external visiting inspection activities.

Quality plans are a requirement of most of the quality assurance standards, e.g. ASME III, BS 5750 Part 1, and they are designed to

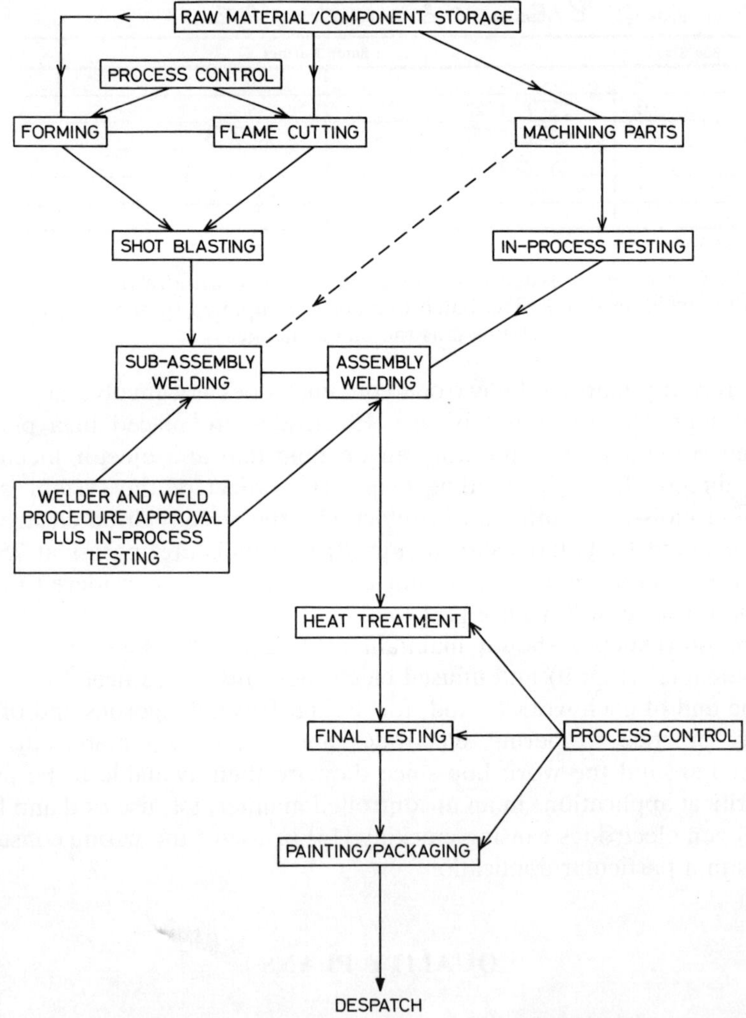

FIG. 10. An outline manufacturing sequence for a steel fabrication showing some elements to be covered by a quality plan.

provide not only a schedule of operations to be carried out by process and inspection personnel, but also to serve as documentary proof that the operations have been performed. It is normal to require the operator, either as a direct worker or an inspector, to sign a control document indicating completion of a stage of the quality plan. 'Hold points' are established in conjunction with visiting inspectors beyond which work must not proceed without the signature of the authorised inspector; these normally apply to inspection, test and documentation stages and are not usually applied to process operations.

The essential feature of a quality plan is that it should describe (Fig. 10) the manufacturing route, the associated in-house inspection stages, and the hold points agreed with the authorised inspector. For most contracts the quality plan, as agreed with the purchaser or his representative, becomes part of the contractual document. The amount of detail to be included in a quality plan by a fabricator is not well defined [15]. The more the fabricator includes as in-house checking the more the authorised inspector may wish to become involved with mandatory hold points, thus potentially delaying production.

There is no standard layout for a quality plan, each fabricator designing a system to meet his own needs. A typical format as used by a pressure vessel fabricator is shown in Fig. 11. The approach adopted in preparing quality plans must not be too system orientated. Excessive paperwork and unnecessary duplication of activities such as auditing and inspection must be avoided. Quality plans should not be regarded as being sacrosanct and flexibility is required at all times, especially if the fabrication is of a one off or of a non-standard nature.

Quality plans are now in common use in the pressure vessel, nuclear, and heavy structural welding fields, and their use is increasing in many other industries.

WELDING PROCEDURE AND WELDER APPROVALS

The requirement for approval of the competence of welders and determination of the suitability of welding procedures goes back to the earliest days of the welding process. At that time the emphasis was on welder testing, usually by mechanical means together with macro-etched sections, and certificates of competence were issued, often by the inspection bodies. Manufacture at that time included welding test plates, the successful testing of which determined the acceptance of the

	CONTROL CHECK	REFERENCE DOCUMENT	CONTROL CATEGORY			
			C	D	A	M
1	Design and Drawing approval	Client's approval required See Sheet			R	
2	Vendor Control	Only NEI ICL approved Vendors shall be used			R	R
3	Materials and Related Purchase Orders	See Sheet 1	R	R	R	
4						
5	Goods Inwards Inspection	QCD 734	R			R
6	~~Receipt of Fin Bar~~	~~QCD 110~~				
7	Receipt of Material in Bar Form	QCD 112	R			R
8	Receipt of Material in Tube Form	QCD 113	R			R
9	Receipt of Carbon Steel Pressure Plate	QCD 100	R			R
10	Receipt of Carbon Steel Non-Pressure Plate	QCD 111	R			R
11	~~Receipt of Stainless Steel Plate~~	~~QCD 110~~				

FIG. 11. Part of a quality plan used for the manufacture of a boiler steam drum. Items not required are deleted and reference is made to standing procedures.

12	Receipt of Header Shell Material	QCD 107	C	D	A	M
13	Inspection of Incoming Castings	QCD 748	C	D	A	M
14	Receipt of Free Issue Material	QCD 183	C	D	A	M
15			C	D	A	M
16	Production Test Plates	QCD 407 QCD 409 See Sheet 1	C	D	A	M
			R	R		R
17	Weld Procedures	QCD 401 See Sheet 1	C	D	A	M
			R	R	R	R
18	Weld Procedure Qualifications	QCD 403 See Sheet 1	C	D	A	M
				R	R	
19	Welder Qualifications	QCD 404 See Sheet	C	D	A	M
				R	R	
20	Welding Consumable Control	QCD 142 QCD 402 QCD 412 See Sheet	C	D	A	M
			R			
21	Reporting of Production Weld Defects	QCD 410	C	D	A	M
22	Flash Butt Welding	P & M 406	C	D	A	M

KEY TO CONTROL CATEGORY ;

R ; Control Check Applies
C ; Certificate Required
D ; Document Required By Client
A ; Client's Approval Required
M ; Mandatory Check Points beyond which Manufacturer cannot proceed without Q.A. Authorisation

FIG. 11.—*contd.*

fabrication. Whilst this was probably satisfactory up to about the 1950s, the increasing number of welding processes and materials to be joined has dictated a different approach.

The current welding codes, AWS, ASME, BS 5500, etc., all require the welding procedure to be proven *before* any work is commenced [16, 17], and that the work itself is then carried out by welders who have demonstrated their competence in the skills needed to make joints conforming to that procedure. The following is a comparison of the approval systems for welding.

Previous

(a) Tested welder → Production work → Mandatory test plate

Current

(b) Approval of welding procedure → Competent operator → Production work → Optional test plate

The older systems of welding control, being very dependent on approved welders and test plates, meant that the fabricator did not know if the work he was producing was acceptable until the final test plate was examined, and large amounts of work could then be in question. By utilising approved weld procedures, qualified welders and systematic control of manufacturing operations, dependence on test plates is now reduced to optional, special circumstances, e.g. to act as a bench mark to measure for irradiation damage in nuclear plant.

The concept of approved procedures is straightforward; however, in practice it is not quite so simple due to the large number of variables involved. Thickness and diameter of pipes and tubes, thickness and joint form in plates, nozzles, base materials, welding consumables and weld processes are just a few of the factors to be considered. Clearly, many procedures are similar and do not require individual proving, and there is therefore an incentive to group or range materials/processes/sizes to provide acceptable results from a single test. A complex test is expensive, for example a $\frac{1}{2}$Cr $\frac{1}{2}$Mo $\frac{1}{4}$V pipe test weld can easily cost in excess of £1500 to make and test to BS 4870, so no one wishes to make more tests than are required.

The code writers of ASME and British Standards have tackled this problem by developing the concept of essential, supplementary variables and non-essential variables, changes which determine if requalification of procedures is required. In ASME IX variables to welding procedures are classified as essential, non-essential and supplementary essential. Essential variables define limits within which the variables can be changed without requalification, but the changes must be recorded by a revision of the welding procedure. Similarly, changes to non-essential variables must be recorded on the welding procedure. The supplementary essential variables define restrictive limits for cases when the base material has guaranteed impact properties. An example of how these principles operate is shown in Table 3 for submerged arc

TABLE 3

COMPARISON OF PROCEDURAL VARIABLES FOR SUBMERGED ARC WELDING AS DEFINED BY ASME IX AND BS 4870

Procedural variables		*ASME IX*			*BS 4870*	
		Essential	*Non-essential*	*Supplementary*	*Essential*	*Non-essential*
Joints:	type of groove		×			×
	change in backing		×		×	
	root gap		×		×	
Base material:	thickness	×		×	×	
	P-group	×		×	×	
Filler material:	F-number	×			×	
	A-number	×			×	
	flux classification	×			×	
	flux addition	×				
	SFA specification	×			} by agreement	
	additional filler	×		×		
Position:	change of position		×		×	
Preheat:	decrease >56°C	×			×	
	post-weld heating		×		×	
	interpass temperature			×	×	
PWHT:	change	×		×	×	
Electrical:	current		×	×	×	
Technique:	deposit size			×	×	
	stringer/weave		×		×	
	cleaning		×			×
	back grooving		×		×	
	oscillation		×	×		×
	stick-out		×			×
	number of passes		×	×		×
	number of wires		×	×	×	
	non-metallic backing		×		×	
	automation		×			×
	peening		×			×

TABLE 4

SOME EXAMPLES OF COMMON *P* NUMBERS FOR STEELS AS USED IN ASME IX

p *number and Group*[a]	*Type of material and typical compositions covered*[b]	*Inclusive Range of specified tensile strength* (ksi)	*Specific examples*
*P*1			
Group 1	Carbon, e.g. C–Mn, C–Si, C–Mn–Si	Up to 65	SA 285 Gr.C SA 516 Gr.60
Group 2	Carbon, e.g. C–Mn, C–Si, C–Mn–Si	70–75	SA 515 Gr.70 SA 738
Group 3	Carbon, e.g. C–Mn, C–Mn–V–N	80–95	SA 737 Gr.C
*P*3			
Group 1	Low Alloy, e.g. C $\frac{1}{2}$Mo, $\frac{1}{2}$Cr $\frac{1}{2}$Mo	55–65	SA 213 Gr.T.2
Group 2	Low Alloy, e.g. Mn $\frac{1}{2}$Mo, $\frac{1}{2}$Cr $\frac{1}{2}$Mo, C $\frac{1}{2}$Mo	55–75	SA 387 GR.2
Group 3	Low Alloy, e.g. Mn $\frac{1}{2}$Mo, Mn $\frac{1}{2}$Mo $\frac{3}{4}$Ni, $\frac{1}{2}$Mo V	80–90	SA 302 Gr.B.
*P*4			
Group 1	Low Alloy, up to 2Cr 1Mo	55–85	SA 213 T11
*P*5			
Group 1	Low Alloy, up to 3Cr 1Mo	60–75	SA 336 F21
Group 2	Low Alloy, up to 9Cr 1Mo	60–100	SA 213 T9
*P*6			
Group 1	Ferritic stainless, up to 13Cr including 11Cr Ti	60–70	SA 240–410
Group 2	Ferritic stainless, up to 15Cr	60–65	SA 249–429
Group 3	Ferritic stainless, 13Cr	85–110	SA 182-Gr.F6b
Group 4	Ferritic stainless, 13Cr 4Ni	110	SA 182-F6NM
*P*7			
Groups 1-2-3	Ferritic stainless, up to 18Cr 2Mo Ti	60–70	SA 268 TP430
*P*8			
Group 1	Austenitic stainless, 19Cr maximum	65–80	SA 213-TP347
Group 2	Austenitic stainless, 25Cr maximum	65–80	SA 312-TP310
Group 3	Higher tensile austenitic stainless	75–120	SA 249-TPXM19
*P*9A	Ni steels, 2$\frac{1}{2}$Ni	63–70	SA 334 Gr.7
*P*9B	Ni steels, 3$\frac{1}{2}$Ni	65–70	SA 203 Gr.D

[a] Specific impact test requirements are associated in Group numbers, within the *P* numbers.

[b] Consult ASME IX for the precise permitted compositions.

TABLE 5
GROUPING SYSTEM FOR STEELS—BS 4870 PART 1

Group	*Type of steel*	*Material grade in BS 5500 (for information)*
A1	C/C–Mn steel with minimum tensile strength in the specification up to and including 430 N/mm^2	M0 and M1
A2	C/C–Mn steel with minimum tensile strength in the specification over 430 N/mm^2	M1
B	C–Mo steel	M2
C	Mo–B steel and Mn–(Ni)–Cr–Mo–V steel	M3 and M4
D	$1/1\frac{1}{4}$Cr$\frac{1}{2}$Mo steel	M7
E1	2(3)Cr 1Mo steel normalised and tempered	M9
E2	2(3)Cr 1Mo steel quenched and tempered	M9
F	$\frac{1}{2}$Cr $\frac{1}{2}$Mo $\frac{1}{4}$V steel	M8
G	5Cr $\frac{1}{2}$Mo steel	M10
H	9Cr 1Mo steel	M11
J	12Cr Mo V steel	M12
K	$3\frac{1}{2}$Ni steel	M5
L	9Ni steel	M6
M	13Cr ferritic stainless steel	—
N	17/20Cr ferritic stainless steel	—
P–Q	Reserved for future allocation of other steel groups	—
R	304 type austenitic stainless steel	—
S	310 type austenitic stainless steel	—
T	316 type austenitic stainless steel	—
U	321/347 type austenitic stainless steel	—
V–Z	Reserved for future allocation of other steel groups	—

welds, made both to ASME IX and BS 4870. Material groupings applicable to these codes are shown in Tables 4 and 5. It will be seen that there is a close similarity between the two codes, although sub-grouping within the ASME system is more specifically designed to accommodate materials having nominated impact values. The groups are the best endeavour to associate base materials having similar weldability composition and mechanical properties. If used intelligently, this reduces the number of qualification tests required.

All the procedure parameters must be accurately recorded at the time of testing and, once documented, the procedure remains valid provided none of the essential variables are changed. The mechanical test requirements of ASME IX and BS 4870 are summarised and compared in Table 6.

TABLE 6

COMPARISON OF MECHANICAL TEST REQUIREMENTS OF ASME IX AND BS 4870 FOR PLATE BUTT WELDS AND BRANCH ATTACHMENTS

Number of samples	*Weld butt*		*Branch attachment weld*	
	BS 4870	*ASME IX*	*BS 4870*	*ASME IX*
Macro-examination	1	—	4	Not required by code
Hardness survey	1	—	1	
Transverse tensile	1	2	—	
All-weld tensile	1	—	—	
Root bend {<10 mm BS 4870; <9 mm ASME IX}	1	2	—	
Face bend {<18 mm ASME IX; <10 mm BS 4870}	1	2	—	
Side bend {>10 mm BS 4870; >9 mm ASME IX}	2	4	—	

Planning of weld procedure testing is essential if the minimum number of tests is to be carried out and qualified prior to production welding commencing. A sequence of events would be:

(a) List all welds involved.
(b) Determine date when they must be completed.
(c) Decide the type of welding procedures to be used.
(d) Classify (a) into groups which can be qualified by a single test.
(e) Set the conditions of the test such that all essential and supplementary essential variables for all welds are covered.
(f) Establish schedule of operations to meet completion date (b).

A different approach is however adopted by the American Welding Society's Structural Welding Code AWS D.1.1. where qualification by individual fabricators of certain common joints and consumables used with any listed steels is not required provided nominated requirements are met. Such welds are designated as 'prequalified' and the code defines the joint configurations, with tolerances, which may be used with manual metal arc, submerged arc, gas metal arc (except dip transfer process), or flux cored arc welding, as appropriate.

Welder approval is based on the operator producing a test piece using parameters determined during the procedure test, but the test is now aimed at demonstrating the welder's *skill* in performing the operation—his ability to produce a sound weld, *not* to reapprove the procedure by a full testing schedule. Variations to the welder qualification procedure are acceptable which would be considered vital to welding procedure qualification. For example, steels and electrode types are grouped together and proof of the welder's skill with one steel or electrode will be acceptable for many others. The basic test positions recognised by the ASME Code are illustrated in Fig. 12 whilst Table 7 shows how the increasing difficulty of certain positions qualifies the welder for other positions which are considered easier to weld. A similar system of qualification positions is given in BS 4871.

Certification of welder approval is valid for a nominated period, three months, six months or one year, but at these intervals checks are made, normally by the employer, to confirm the continued skill of the operator. If any serious deterioration of capability occurs as shown for example by poor non-destructive testing performance, the operator will be advised and subjected to retraining and/or recertification. If the record is satisfactory, the certificate is endorsed and is valid for a

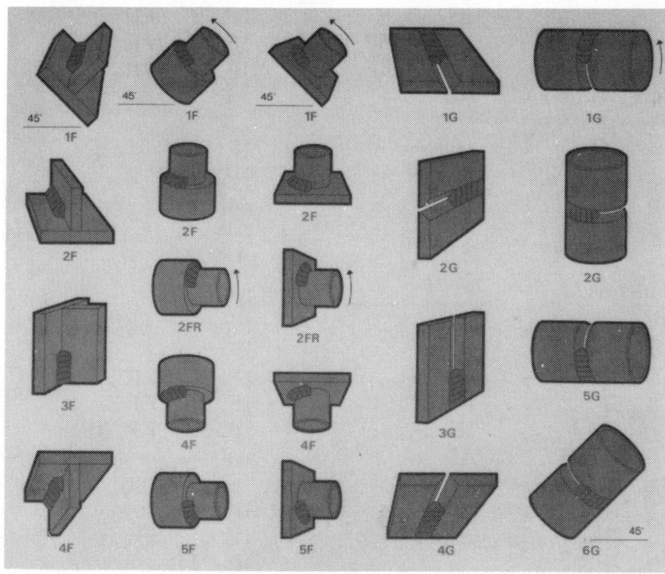

FIG. 12. The basic test positions for fillet and groove welds, as defined by ASME IX.

TABLE 7
ASME IX PERFORMANCE QUALIFICATION AND ASSOCIATED LIMITING POSITIONS

Qualification test	*Position and weld type qualified by test*	
Butt weld	*Butt welds*	
Plate position	*Plate*	*Pipe*
1G	1G	1G
2G	1G, 2G	1G, 2G
3G	1G, 3G	1G
4G	1G, 4G	1G
3G, 4G	1G, 3G, 4G	1G
Pipe position		
1G	1G	1G
2G	1G, 2G	1G, 2G
5G	1G, 3G, 4G	1G, 5G
6G	All	All
2G, 5G	All	All
Butt weld	*Fillet welds*	
Plate position	*Plate*	*Pipe*
1G	1F	1F
2G	1F, 2F	1F, 2F, 2FR
3G	1F, 2F, 3F	1F, 2F, 2FR
4G	1F, 2F, 4F	1F, 2F, 2FR, 4F
3G, 4G	All	All
Pipe position		
1G	1F	1F
2G	1F, 2F	1F, 2F, 2FR
5G	1F, 2F, 3F, 4F	All
6G	All	All
2G, 5G	All	All
Fillet weld	*Fillet welds*	
Plate position	*Plate*	*Pipe*
1F	1F	1F
2F	1F, 2F	1F, 2F, 2FR
3F	1F, 2F, 3F	1F, 2F, 2FR
4F	1F, 2F, 4F	1F, 2F, 2FR, 4F
3F, 4F	All	All
Pipe position		
1F	1F	1F
2F	1F, 2F	1F, 2F, 2FR
2FR		1F, 2FR
4F	1F, 2F, 4F	1F, 2F, 2FR, 4F
5F	All	All

further period. Normally a certificate lapses if no work of the type covered by the certificate has been performed in the preceding period.

CALIBRATION OF WELDING EQUIPMENT

Calibration of welding equipment and associated measuring devices is commonly subject to investigation during a quality audit and this normally presents few problems, in a well organised works, other than the establishment of an effective system, and the discipline associated with implementation. However, problems are often encountered with drying and baking ovens for electrodes, particularly at higher temperatures, above 300°C, when the temperature may be shown on the recorder but not achieved in electrodes. (Failure to achieve the required temperature may also be a function of the way the electrodes are loaded into the oven (Fig. 13) since, if the layers of electrodes are too thick, heat may not reach to the centre of the bundle in the nominated time [18].

Preheating and post-weld heat treatment of welds is critical to the success of the welding operation. Preheat is often measured by temperature-sensitive crayons which, whilst generally accurate in themselves, do not provide a continuous reading of temperature. Portable and permanent devices for measuring preheat, including pyrometers, digital devices, etc., should be subjected to regular (not exceeding three months) calibration against a known standard and should be marked with a label showing the date of the next calibration, in addition to the normal book records. Furnaces used for heat treatment normally contain permanent thermocouples recording the furnace atmosphere temperature, whilst portable thermocouples are often attached to measure the actual *metal* temperature on a multi-channel recorder. Both these systems require regular checking and certification for conformity.

It is difficult to give guidance as to what calibration practices should be adopted for welding equipment itself. Nicholson and Cresswell [19] have reviewed the problem and conclude that where the operator is himself responsible for precise control of the process, e.g. manual metal arc, manual TIG or MIG, maintenance of equipment is probably more important than calibration. In more sophisticated processes, e.g. automated TIG, equipment design reduces the inherent variables

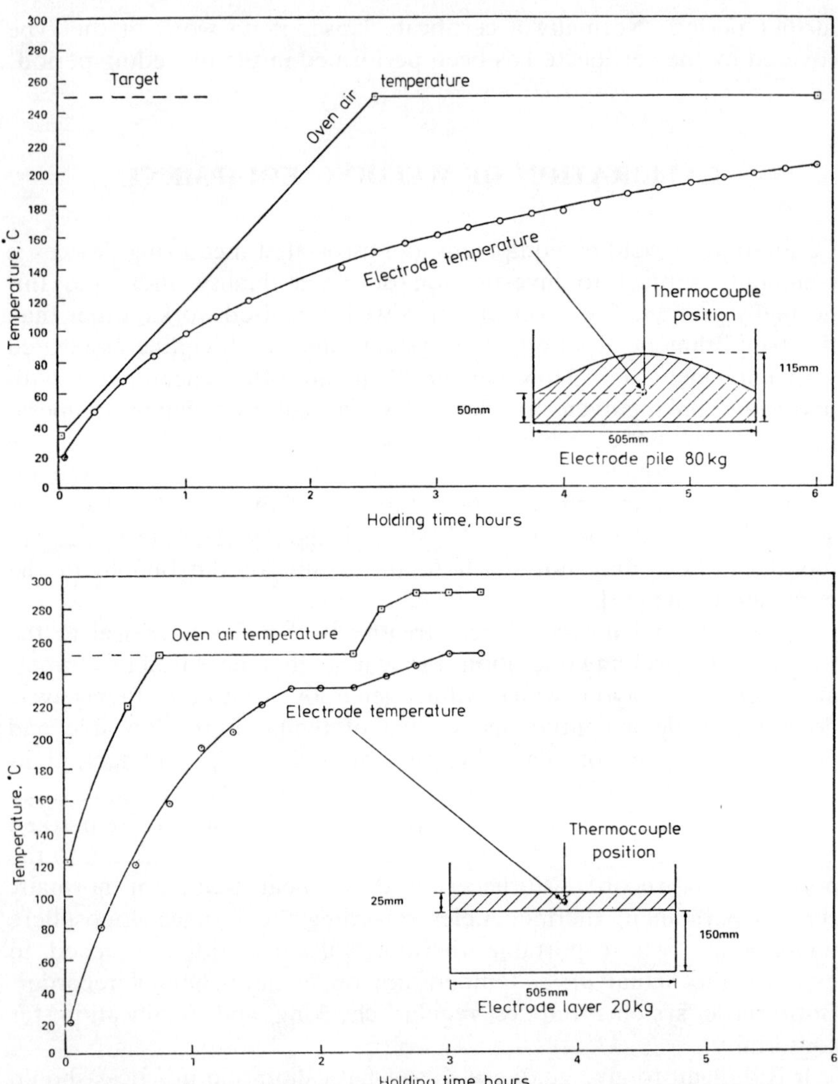

FIG. 13. The effect of loading electrodes into a drying oven in piles or bundles rather than thin layers of *c.* 25 mm showing how readily electrodes can be inadequately dried. Top—temperature rise after loading 80 kg of electrodes into a hot oven at 250°C original temperature. Bottom—temperature rise after loading 20 kg of electrodes into a hot oven at 250°C original temperature. (After Boniszewski and Pavely, reference 18.)

which the operator can influence and calibration becomes more significant and should be subject to routine checking against primary standards.

WELDING OPERATIONS

There is a series of distinctive actions which are required to be performed prior to, during, and subsequent to the welding operation, apart from systematic activities described elsewhere in this chapter, e.g. consumable control. Not every action is required for every construction, but a conscious decision should be taken when omitting any activity, e.g. not providing local environmental protection. The major requirement is to ensure that the qualified weld procedure is followed as accurately as possible.

Prior to Welding

(1) *Protection* of the welder from the elements such that the adopted welding process is not adversely affected. In some cases it may be essential to erect a temporary sheeted cover over the workpiece to prevent the effect of draught or wind when using gas shielded processes such as MIG or TIG, even in a shop environment, apart from the need for protection from the elements when working outdoors.

(2) Care must be taken to ensure *cleanliness* of the work. This may be achieved by wire brushing, grinding, solvents, etc., depending on the nature of the contaminant. Clean materials should not be assembled for long periods prior to welding or dirt or grease may enter into joints and affect weld quality. Care is required to ensure that dirt and rust are not trapped in joints involving fillet constructions or porosity may result.

(3) *Weld preparation* dimensions and *fit-up* should be carefully checked and rejected if not conforming to the welding procedure requirements. Accidental reduction of the size of the weld preparation may give insufficient access to deposit weld metal in the required sequence or by the nominated process (Fig. 14). On the other hand, if the weld preparation is too large, distortion may occur together with excessive shrinkage, apart from incurring additional costs due to the increased volume of weld metal. Similarly; the fit-up of joints requires close attention. Misalignment, mismatch or incorrectly sized gaps can, again, make the joint unacceptable.

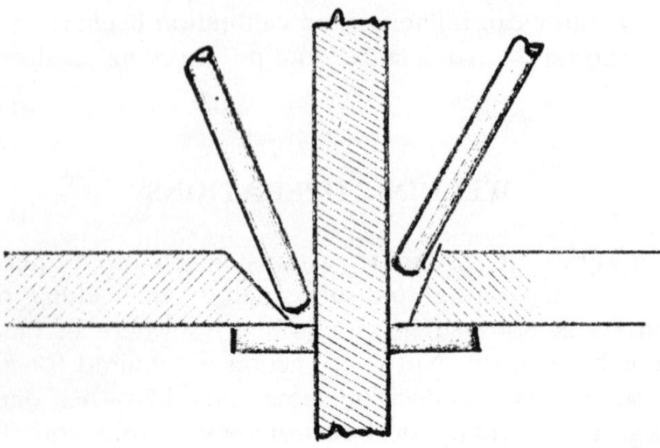

FIG. 14. How the joint weld preparation angle on the root gap can limit access for welding. On the left hand side a 45° angle is adequate to reach the root area but on the right hand side with $22\frac{1}{2}$° the weld metal will bridge without fusing into the root.

(4) *Preheat* metal temperatures should be determined and welds should not be made if the temperature has not been achieved within 75 mm of the welding point. Preheat is sometimes applied by placing the component in a furnace—in such cases the metal temperature of the workpiece must be measured—*not* that of the furnace.

During Welding

(1) *Tack welds* should be examined during welding to ensure they are sound and have not broken. Failure to do this can result in distortion or incorrectly dimensioned fabrications.

(2) *Welding parameters* should be checked, including welding current, polarity, electrode or wire sizes, travel speed and gas flow rates, when applicable.

(3) The *deposition sequence* of weld runs should be confirmed to ensure that no pockets are formed and that deslagging is effective (when required). Control of distortion techniques must be observed [20].

(4) *Preheat* and *interpass* temperatures should be constantly monitored. In some cases the work must not fall below specified tempera-

tures and in others must not exceed a given temperature prior to deposition and subsequent run of weld metal.

(5) In-process *non-destructive testing* should be applied at any initial stages of welding, e.g. magnetic particle testing of root passes, as required by the quality plan.

(6) When required by the contract, control of the *weld consumables* should be maintained to each welder concerned in the work and joints or parts of joints should be identified as being the work of specific welders. This requires careful control of the welders to ensure that they continue to operate in their designated zone.

After Welding

(1) The final *geometry* of the weld should be examined for size and shape, undercut, porosity and cracks.

(2) Welds which are ground should be examined for excessive removal of metal which may reduce the effective thickness of the component.

(3) Particular attention should be paid to the removal of spatter and temporary assembly cleats.

(4) Non-destructive testing should be carried out by competent operators. Care should be taken in certain cases, e.g. high strength low alloy steels, that the prescribed period (normally about 48 h) has elapsed before testing is permitted to commence, since delayed hydrogen cracking may occur.

(5) Repairs to any defective areas should be made to approved procedures, the workpiece having been clearly marked in an unambiguous manner to prevent the repair being carried out at the wrong place.

(6) Heat treatment, when required, should be confirmed as conforming to the procedure in respect of temperature gradients, heating and cooling rates, location of thermocouples, etc.

WELD RECORDS AND FEEDBACK OF INFORMATION

In a number of cases the fabricator is required by the code of practice to maintain records of some aspects of welded constructions. These records are normally provided to the purchaser and may even form part of the contractual agreement. Such records commonly include

[21]:

(a) material test certificates;
(b) weld procedure test records;
(c) welder performance certificates;
(d) heat treatment charts;
(e) reports on non-destructive testing;
(f) hydraulic test certificates.

In the case of complex or very large projects, the compilation of the required data can require a significant effort on the part of the fabricator, and a systematic approach, based on the quality plan, is required to ensure all data is input to a central source from all individual activities engaged on the project, for example, metallurgy, inspection, NDT, etc. The use of in-house, pro-forma documents is to be commended to ensure that all the required information is provided. The amount of paperwork required can be significant and Fig. 15 shows the manual compilation of data books for a large steam drum. The use of microfilming of data is now common and it would be of

FIG. 15. Compilation of data books for the fabrication of a high quality steam drum.

great value in many cases if purchasers would accept these in lieu of the paperwork.

There are few cases where the codes require records of the use of welding consumables, although many fabricators find it beneficial to be able to identify which consumables have been used in which seams. Welders can be issued with seam record cards (Fig. 16) when they are given the instruction on weld procedures, and the completed cards can be controlled by the welding engineer. Records of weld quality are often maintained by fabricators in order to be able to measure the effectiveness of their workshops and welders (Fig. 17). Such data is normally compiled by the inspection and non-destructive testing departments and is provided at regular intervals of, say, a week or a month, to senior production management.

There are a number of ways in which the information can be supplied in terms of weld seam length, sub-components welded or items completed, the amount of satisfactory work being measured against that found to be defective [22], i.e.,

$$\text{rework } (\%) = \frac{\text{defective weld (mm)}}{\text{total weld (mm)}} \times 100.$$

This form of measurement gives a quick overall reading of how an individual welder or workshop is performing; the shop manager or welding engineer soon learns if he normally operates at, say, '3 % defective' and can take prompt action to deal with any situation if a drift is noted. It also enables a measure to be made of the effectiveness of changes, i.e. introduction of a new welding process.

A major weakness is however when the product mix is variable, especially in respect to the thickness of material welded, and using the above method the performance on 20 mm thick plate makes the same contribution to the rework return as, say, 100 mm thick welds containing six or eight times more weld metal.

In this case it may be found more valuable to use a unit length/thickness criteria. For example, it is possible to use units of, say, 100 mm length and 10 mm thickness, to describe the total amount of weld employed in a particular joint and relate this to any unacceptable quality. A butt joint in, for example, 110 mm thick plate, would then rate as 11 units for thickness and, if 1000 mm long as:

$$11 \text{ thickness units} \times \frac{1000}{100} \text{ length units} = 110 \text{ units}.$$

NEI International Combustion Ltd
WELDING RECORD SHEET

REF. QCD 408

PROC. NO. KC/296/4.
W.Q. NO. 814.

CONTRACT NAME: N W CORDER.
CONTRACT NO. JO494.
VESSEL NO. 2 SEAM 4.
ISSUE DATE 15/11/80.

ORDER NO. B 421/1106/001.
MATERIAL BS4360-50D.

WELD SKETCH

Name	Pressure No.	Weld Identifi-cation	Consumables	Type	Gauge	Batch No.	No. of runs	Amps	Volts	Speed mm/min
WRIGHT.V.	41	SIDE 1.	EAB 55.00	–	3·2	131260	1–2	120	—	—
			ESAB 55.00	–	4	161021	3–4	170	—	—
SMITH.N.	210	SIDE 1.	SD3	–	4	6491	5–19	420	30	400
			BX300	–		321/6				
SMITH.N	210	SIDE 2.	SD3	–	4	6491	1–	450	30	400
			BX300	–		321/6				

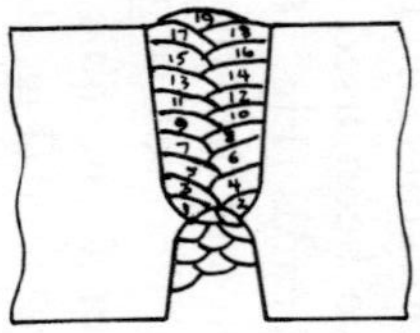

COMMENTS
SLAG TRAP AFTER RUN 12 – GROUND OUT.

SPECIAL REQUIREMENTS
TO BE SOAKED AT 150 °C FOR 2 HOURS ON COMPLETION.

OM 53971/1 | DATE COMPLETED | AUTHORISED SIGNATURE

FIG. 16. A partially completed weld seam record card for a submerged arc weld.

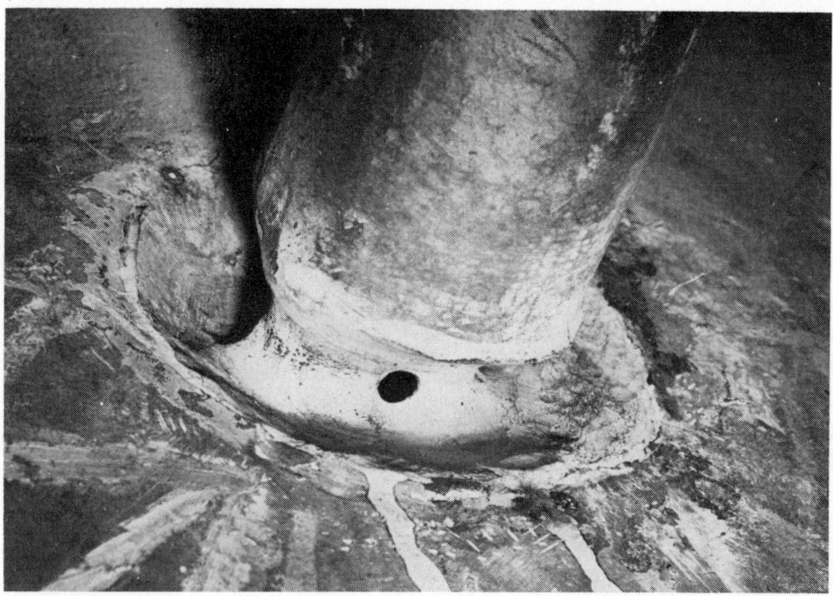

FIG. 17. An extreme example of excavation of a 75 mm diameter stub weld to remove slag, detected by ultrasonic testing and trapped above the argon arc root pass.

If a defect then ran through 2 such units, a measure of overall quality would be seen as $\frac{2}{110}$ or less than 2 % actual defective metal, not 20 % if recorded on the basis of seam length only.

Data can also be kept, normally using a computer based system, of the rework causes and costs which are then segregated by a cause code for subsequent analysis. Such data can be of great use in pinpointing problems to a particular shop, office, machine or process, and enable effective corrective action to be taken. Some examples of the cause codes used by one fabricator are given in Table 8. Given such data and its cost, associated with the overall costs of quality, it is possible to provide senior management with information which enables the efficiency of quality operations to be measured [23].

The feedback of information to the welding engineer and designer from the shop floor, erection staff and operating personnel, is essential if reliable, economic fabrications are to be produced. It should never be too much trouble to report difficulties since, unless reported, they

TABLE 8
EXAMPLE OF A REWORK CAUSE CODE

Code	*Cause*	*Notes*
0031	Drawing errors, route cards, documentation	Incorrect drawings, revised drawings, issued after job has started resulting in scrap or rework; incorrect documentation
0032	Defective, incorrect or lost material	For any reason whether due to internal errors or incorrectly furnished by suppliers
0033	Layout	Material laid out incorrectly, wrong patterns used, etc., marking out; jigging
0034	Material preparation	Burning, grinding operations
0035	Tube manipulation and tube preparations	Unsatisfactory bending, incorrect weld preparations
0036	Defective castings	All causes
0037	Forming and pressing operations	Defective press tools, incomplete bending, incorrect material temperatures
0038	Machining errors due to operator mistakes	
0039	Defective machines	Machining errors caused by inadequacies of the machine
0040	Fitting errors	Including erection fitting, electrical fitting, etc.
0041	Fit up/clean up	Butt welds, nozzles, stubs, lugs, etc., not attached correctly (out of position etc.); scale, slag, spatter, burns and sharp edges not removed
0042	Manual welding (MMA)	Undersize or oversize welds, undercut defective welds, welds not to drawing, incorrect rods, etc.
0043	Welding	All methods except MMA
0044	Heat treatment errors	Including incorrect pre- or post-heat
0045	Handling, packaging, storage	Machine surface damages, damage caused during internal transport, deterioration due to incorrect or poor storage, damage to packaging etc.
0046	First time weld repairs	

TABLE 8—*contd.*

Code	*Cause*	*Notes*
0047	Quality control errors	Incorrect NDT sentencing, incorrect inspection leading to rework, incorrect heat treatment instruction, incorrect quality literature
0048	Miscellaneous	Includes all reasons not listed above; a full explanation should be made in the rejection notice
0049	Site damage/losses	All work connected with site

will repeat themselves in other circumstances and inevitably cost money to correct.

EXAMPLES OF PROBLEMS ASSOCIATED WITH WELDING

The earlier part of this chapter described a sequence of actions which, if correctly followed through the manufacturing cycle of a fabrication, will largely guarantee a successful product. We do not however live in a perfect world and a number of cases are now given which illustrate the consequences of failure to observe these quality principles.

Test Plate Failure

The test plate for a steam drum was welded in 102 mm thick BS 1501-223-32B, by submerged arc welding using SD3 wire and BX 200 flux. The plate was examined by X-ray and subsequently heat treated in accordance with BS 1113 at 600°C. The minimum tensile strength of the joint was required to be 494 N/mm^2, and all weld test pieces were taken from the top and bottom sides of the test plate. The results obtained were—top of the weld, 457 (UTS) N/mm^2; bottom of the weld, 512 (UTS) N/mm^2;—representing a shortfall of 37 N/mm^2 for the top tensile test piece.

Comprehensive examination of records of consumables used showed no identifiable problems and analysis of the tensile test pieces gave the following results:

	C %	Si %	Mn %	P %	S %	Cr %	Mo %	Ni %	Cu %
Top:	0·07	0·26	1·61	0·017	0·007	0·05	0·01	0·04	0·27
Bottom:	0·06	0·28	1·62	0·014	0·010	0·04	0·01	0·04	0·30

These figures clearly indicate that there was no error in the consumables used. Hardness tests on the tensile samples showed average values of 206 Hv_{10} at the top and 215 Hv_{10} at the bottom. Macro- and microexaminations indicated that, whilst the lower sample consisted of alternate unrefined acicular ferrite and refined zones of fine-grained polygon-ferrite, the latter being caused by the normalising and tempering action of successive weld passes, the upper sample contained a disproportionate amount of refined zone (see Fig. 18). This would account for the lower tensile strength recorded and probably arose from the weld bead deposition sequence, which differed from top to bottom of the weld, coupled with an excessive interpass temperature.

The Lesson

Even in a well executed weld, using proven consumables, the weld sequence and/or the interpass temperature can result in unsatisfactory mechanical properties.

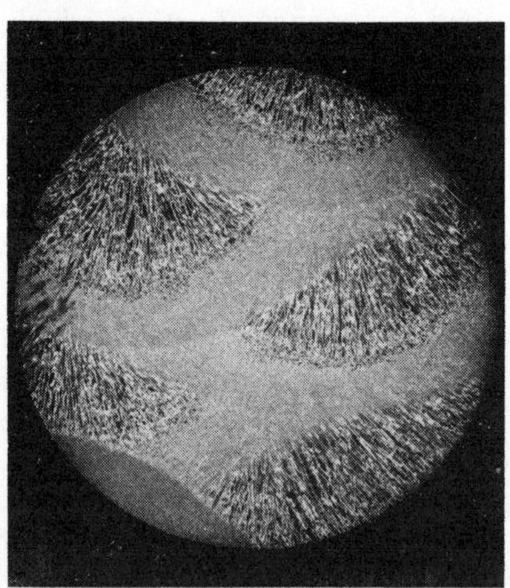
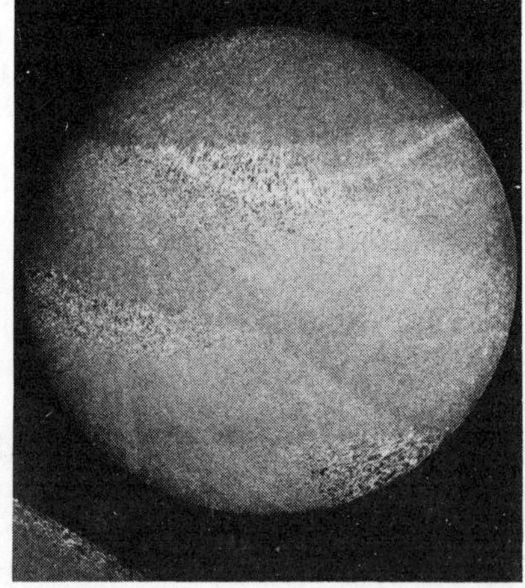

FIG. 18. The carbon–manganese submerged arc weld on the left hand side with clear, alternate refined and unrefined ferrite, gave 512 N/mm^2 UTS whilst the essentially refined weld metal on the right hand side from the same weld, gave only 457 N/mm^2 UTS.

Yoke Failure on Bending Press

A large, 200 ton, vertical plate bending machine was closed by a 400 mm thick carbon steel hinge, after the plate was inserted between the joining rolls. During initial trials of the machine one winter's morning the yoke piece broke clean through in a brittle manner. Examination showed that the failure originated from an 8 mm intermittent fillet weld, attaching a guide member on the inside of the yoke (see Fig. 19). The weld had been made using rutile electrodes, with no preheat, onto a flame cut surface, and was not subjected to heat treatment.

The Lesson

Small welds on highly stressed members are very prone to cracking, if inadequate welding procedures are used, thus initiating failure. Such welds should be avoided as far as possible. In this particular case, on the replacement unit, the guide member was attached by setscrews!

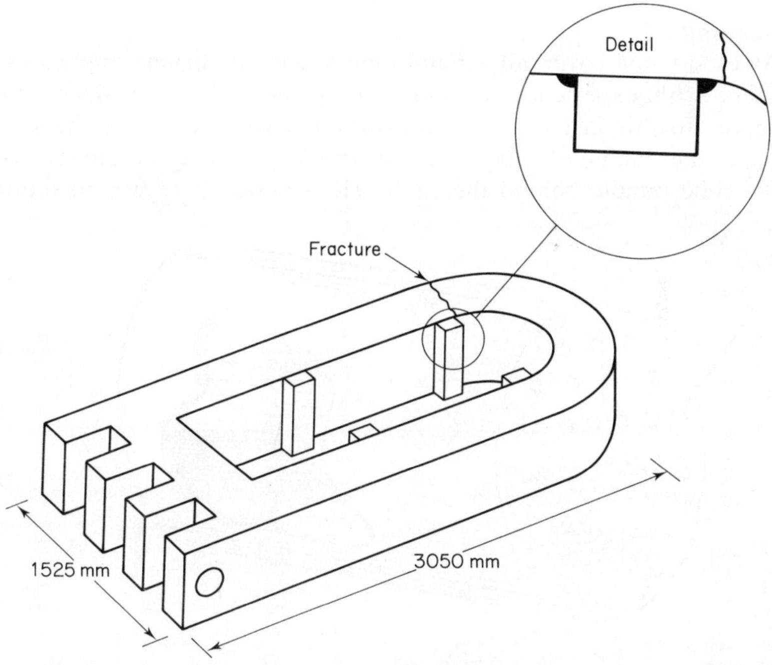

FIG. 19. Failure of yoke piece from a 200 ton vertical plate bender. The origin of the fracture was from an intermittent fillet weld attaching a guide bar.

Furnace Heat Treatment of Waste Heat Boiler

A medium sized waste heat boiler measuring about 4400 mm long and 2290 mm diameter, was made from 56 mm carbon steel to BS 1501-151-28A. The unit had flat ends, 45 mm thick, through which 500 carbon steel tubes, 44·5 mm × 5 SWG, were expanded and seal welded, and into which 14 large diameter stay bars were welded (Fig. 20). On completion of fabrication the unit was furnace stress relieved, as required by the code, at 600°C. The metal temperatures were measured on the end plates and on the shell, with the axis of the vessel horizontal. When the vessel had cooled it was found that the outer ring of tubes had bowed by as much as half-diameter, and others to a lesser extent, due to permanent elongation of the tubes.

Subsequent investigations showed that the tubes in the centre of the bundle must have been up to 100°C colder than the shell and outer tubes during the heating cycle, leading to significant residual stress and resulting in the deformation of the tubes.

The Lesson

Codes do *not* cover all circumstances and intelligent application, together with experience, is required by the welding engineer and designer. In this instance the fabricator failed to recognise the mass effect of the bundle of tubes and the consequential lag in temperature of the tube bundle behind the shell. The solution here was to retube

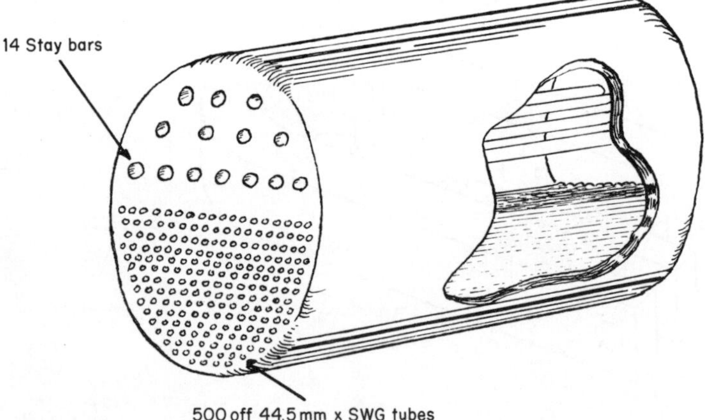

FIG. 20. Cut-away view of a waste heat boiler showing tube bundle and stay bars.

and reheat-treat with the vessel vertical, using thermocouples inside the tube bundle to control the heating and cooling rates, having established that a further stress relief treatment would not lower the properties of the shell plate below the design minimum [24].

Stainless Steel Tube Failure

It is sometimes a requirement for butt welds in stainless steel tubing to be locally solution treated after non-destructive testing has been carried out. In a power station a number of SA 213 TP 316 tubes, 27·5 mm OD × 12 SWG, were required to be heat treated at 1030 to 1060°C for 30 min, followed by rapid removal of insulation and the electric heater. The temperature of each weld was measured using a thermocouple attached by capacitor discharge welding, the leg of which was bound with wire to the tube, to avoid damage to the thermocouple.

When cold a number of tubes broke transversely, just clear of the weld, in a brittle manner, and displayed an oxidised crack face. A light tan coloured deposit was noticed on the fracture face, and analysis revealed that this contained large quantities of zinc. Detailed investigations showed that the failure was due to hot shortness (of the material in the presence of molten zinc) and arose directly from binding of the thermocouple to the tube with galvanised wire, when heated at 800 to 950°C, and under a stress of 3000 to 6000 psi.

The Lesson

Great care must be taken at all times to ensure that incompatible materials do not come into contact with susceptible welded fabrications. A further example, commonly encountered, is when the copper alloy contact tips on submerged arc welding heads are accidentally in contact with the weld pool—cracking will inevitably result.

Mechanical Properties of a Submerged Arc Weld

An SD3 $\frac{1}{2}$Mo wire used with an acid flux had satisfied all requirements for the welding of a low alloy steel (Mn, Cr, Mo, V) with tensile strength of 560 N/mm^2 (min) and a yield strength of 385 N/mm^2 (min). During a routine examination of a test plate the following all weld tensiles were recorded:

	Specimen 1	Specimen 2
Yield stress (N/mm^2)	388	404
UTS (N/mm^2)	492	532

The UTS values do not meet the minimum specified requirement, and a full investigation ensued. The wire, weld deposit, and the wire used for the initial tests were analysed, with the following results:

	C %	Mn %	Cr %	Mo %
Original wire	0·154	1·91	0·10	0·621
Production wire	0·11	1·78	0·10	0·47
Weld deposit	0·04	1·11	0·15	0·50
Wire specification	0·16 maximum	1·60/2·20	—	0·40/0·70

The inferior properties are clearly due to the lean analysis of the weld deposit, with carbon and manganese being particularly low. The carbon loss of 0·07 % was considered to be particularly significant and was ascribed to the use of the acidic flux. The short term solution to the problem was to control the wire to 0·15 % C minimum, 1·8 % Mn minimum, and 0·50 % Mo minimum, whilst in the longer term a different combination of wire and flux (basic) was adopted.

The Lesson

Welding consumables are supplied within a range of compositions and care must be taken to ensure that procedure test results are not obtained from wire at the upper end of the range, or unsatisfactory mechanical test values may be obtained during production from materials at the lower end of the specification. The fact that the wire specification had a maximum carbon value but no minimum was significant, particularly when used with an oxidising flux.

Brittle Failure of Large Water Turbine Spiral Casing

A turbine spiral casing, some 2·4 m diameter, was fabricated in a 'lobster back' manner from 42 mm high-tensile carbon steel conforming to the now obsolete BS 968:1962. The specified composition was (all maximum values):

C %	Si %	Mn %	S %	P %	Cr %	Ni %	Cu %
0·23	0·35	1·8	0·060	0·060	0·35	0·50	0·60

A branch pipe, some 1000 mm diameter, was attached at an angle of about 45° to the axis of the spiral, compensation being provided for the opening by a large reinforcing ring. Welding was by manual metallic arc using low hydrogen electrodes and with a nominal preheat of 100°C. Non-destructive testing was by radiography of main seams, and magnetic particle inspection of fillet welds. Cracking problems were

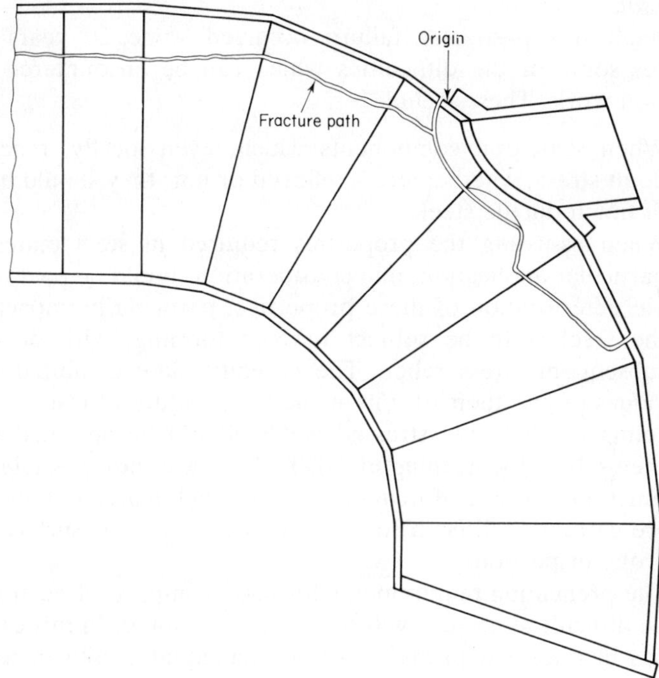

FIG. 21. Sectional plan view of failure of water turbine spiral casing showing path of crack.

encountered during fabrication in a number of welds and the unit was given an intermediate stress relief at 600°C (metal temperature) prior to a final stress relief at 625°C. The cause of this cracking was not determined.

The unit was undergoing hydraulic testing at 10°C (50°F) to achieve a design pressure of 53 bar, when the spiral burst in a brittle manner at 52 bar. The fracture ran from the edge of the compensating ring in both directions and was some 8 m in length (see Fig. 21). Examination showed the plate to possess very poor impact strength, although this property had not been specified when the plate was ordered. There was a large, almost through-thickness, repair weld, some 1·3 m long, in the spiral plate adjacent to the compensating ring fillet weld, in which extensive heat affected zone cracking had occurred, and which acted as the origin of the crack. The particular plate was severely segregated, a factor which probably contributed to the cracking.

The Lesson

Although this particular failure occurred some 20 years ago it illustrates some of the difficulties which can be encountered during fabrication work. These include:

(a) When structural components which, even locally, reach yield point stress, whether stress relieved or not, they should be made of notch ductile steel.

(b) When assessing the properties required in steel plates for a particular application, due consideration should be paid to possible deterioration of these properties, particularly impact, when the steel is to be subject to cold forming, with or without subsequent stress relief. The casualty plate exhibited Charpy values of less then 10 ft/lb at the temperature of test.

(c) Joints in all highly stressed welds should be designed to lend themselves to meaningful NDT. This was not possible at the junction of the reinforcing ring to the shell plate, and fillet welds are extremely difficult to examine, other than by surface microprobe inspection.

(d) The preheating requirements for large complex fabrications may be difficult to achieve without careful planning. In this case local gas heating was probably totally inadequate, both in temperature reached and in its distribution.

(e) All major repairs to welds, or fabrication difficulties, should be reported to a central authority, by shop management, to determine any additional precautions to be taken. The incidence of cracking during manufacture of the spiral was possibly due to the severe segregation which contributed to the eventual failure.

(f) Weld repairs should be rigorously controlled to conform to approved weld procedures and should be subject to NDT at least as thorough as required on original welds. In the case in question the weld repair was not made to an agreed procedure and no records of NDT of the repair were available.

Effect of Base Material Manufacturing Route

Manufacture of collector and distributor headers for power stations, in a carbon–manganese steel, with a tensile strength of 32 tons/in^2, had proceeded without difficulty for a number of years [25]. The main header body was about 380 mm diameter and 50 to 70 mm thick, and consisted of an assembly of forged 'T' pieces and elbows, joined by wrought pipe, made by full penetration butt welds, and to which was added many tube stubs, 50 mm outer diameter × 8 mm wall thickness (Fig. 22).

The basic manufacturing procedure involved a preheat of 100°C, the use of low hydrogen electrodes, together with radiography of butt welds, and 10 % magnetic particle testing of stub welds, before and after stress relief.

There had been no incidence of cracking when, without warning, cracks were found at the toes of an unprecedented number of stub welds (see Fig. 23) on forged elbows and 'T' pieces. After much heart searching and a process of elimination, it was concluded that the welding procedure was not at fault and the forging material was abnormally crack-sensitive. This conclusion was supported by the fact that components fabricated from material from another supplier were crack-free.

The cast analysis of the forgings was critically examined and no differences of significant consideration were noted. The analysis is as follows:

	Cracked forgings	Sound forgings
Carbon, %	0·22	0·23
Manganese, %	1·10	0·97
Sulphur, %	0·008	0·032
Phosphorus, %	0·041	0·024
Copper, %	0·19	0·07
Nickel, %	0·17	0·13
Tin, %	0·037	0·008
Nitrogen, %	0·011	0·012

There was no difference between the pearlite ferrite distribution, but it was noted that the steel of the cracked forgings was significantly cleaner.

It was established that the cracked material had been made by a basic electric arc, oxygen blown, vacuum cast route, whilst the sound forgings were made from acid open arc steel, the substitution having been made by the steelmaker without consultation with the fabricator. It was postulated at the time that the cracking was due to the cleanliness of the basic electric steel forging, and that the low volume of sulphide did not provide a sink into which hydrogen could be temporarily absorbed, thus avoiding incipient cracking. (A subsequent

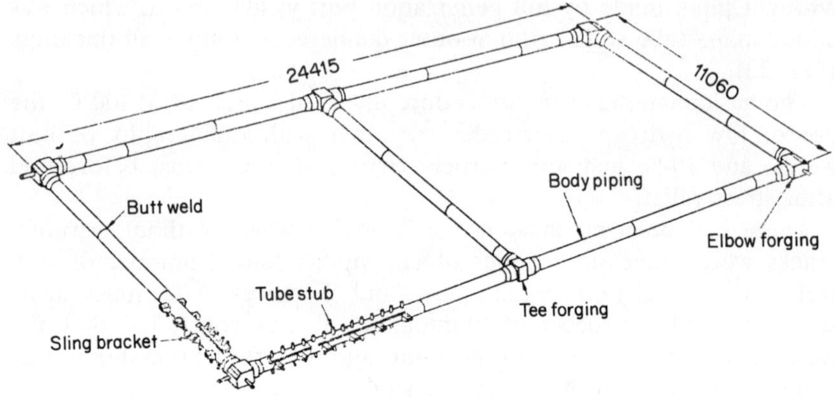

FIG. 22. Typical header assembly for high pressure steam generators.

programme of work at The Welding Institute has suggested that with rolled plate the presence of inclusions helps to nucleate ferrite during cooling and that by reducing the number of inclusions this transformation is inhibited, resulting in a harder structure [26]. This investigation did not include forged products.)

Production was resumed using the same welding procedure but introducing a long post-weld hydrogen release treatment prior to stress relief, which proved effective in preventing cracking.

The Lessons

(1) Even a valid and well proven weld procedure can fail if the base metal is variable.

(2) Close liaison must be maintained between the user and the steelmaker, since changes in manufacturing routes of raw materials may have significant effects on fabrication.

(3) The user should indicate to the steelmaker the intended use of the material and any major working to be performed (e.g. cold spinning) since the latter can often advise on its probable effectiveness.

(4) The fact that the material was 'better' in respect to cleanliness was not so in respect to its weldability.

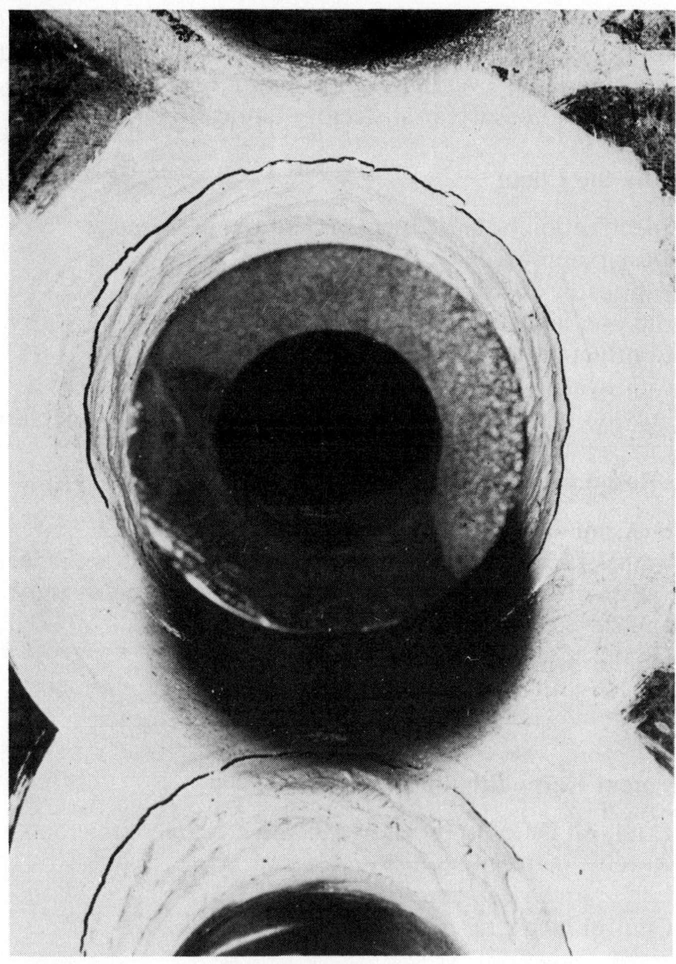

Fig. 23. Typical weld toe cracking as shown by magnetic crack detection techniques, on a low sulphur, carbon–manganese steel forging.

SUMMARY

The preceding sections of this chapter can be summarised into a number of key words which will prove useful aide-memoires. The lists are by no means exhaustive, and certainly do not purport to indicate that the nominated activities must be performed by a single department, or indeed that they are the only actions to be performed. They do however give an overall view of major requirements to be completed in order to ensure a satisfactory fabrication.

Actions by the Client

(a) Specification of basic function and requirements.
(b) Design appraisal.
(c) Fabricator appraisal.
(d) Bid appraisal.
(e) Contract award.
(f) Approval of design, procedures and sub-contractors.
(g) Provide surveillance of manufacturing/inspection operations.

Design Responsibilities

(a) Determine function and material needs.
(b) Translate into working drawings and specifications. Incorporate value engineering, quality engineering, reliability and research and development.
(c) Establish tolerances and acceptance criteria.
(d) Design joints for welding access and testing.
(e) Observe economic and programme requirements.

Procurement Responsibilities

(a) Establish list of approved vendors and sub-contractors.
(b) Receive material requisitions.
(c) Issue purchase orders.
(d) Control incoming materials.

Manufacturing Responsibilities (including Quality Control)

Prior to Welding

(a) Review specification requirements.
(b) Prepare quality plans/inspection points/holds.
(c) Prepare work programme/instructions.
(d) Determine machines and equipment.
(e) Select processes—establish procedures.
(f) Receive and control material, including welding consumables.

(g) Prove welding, heat treatment and NDT procedures.
(h) Approve welders and procedures.
(j) Establish weld consumable control.
(k) Liaise with inspection agencies.

Note standing activities—documented system covering (i) calibration (ii) document change control (iii) work status and (iv) effective stores.

During Fabrication and Welding

(a) Control cutting, assembly, welding. (Measure, visual examination and NDT to specification).
(b) Involve inspection agency.
(c) Control welding consumables.
(d) Repair unacceptable welds.
(e) Make test plates.

After Fabrication

(a) Involve inspection agency.
(b) Perform post-weld heat treatment.
(c) Perform post-weld heat treatment non-destructive testing.
(d) Carry out proof tests.
(e) Approve test plates.
(f) Measure, and final inspection.
(g) Document.
(h) Prepare for shipment.

CONCLUSIONS

There is no bonus for the purchaser or the fabricator if the quality of a welded fabrication exceeds the level required for satisfactory service. Conversely, a defective weld, requiring reworking, adds nothing but cost to the construction. The implementation of a formal quality system will do much to ensure that the required quality is obtained from a welding process, at the first attempt. The key to success of such a system is enthusiastic initiation with subsequent full support from management and a team work attitude by every employee engaged in the construction operation, on the shop floor and in associated activities.

ACKNOWLEDGEMENTS

The author wishes to thank the Directors of Northern Engineering Industries Limited for permission to use some of the data and illustrations. Thanks are also due to my many colleagues for helpful discussions, especially to Mr N. Scrimgeour and Mr Dennis Smith. Assistance from various manufacturers of equipment and materials is also gratefully acknowledged.

REFERENCES

1. Lancaster, J. F. (July 1974). QA through optimum choice of materials. *Petroleum International,* 38–42.
2. Lamellar Tearing in Welded Steel Structures. The Welding Institute, London, 1972, 16 pp.
3. Ott, C. W. (April 1979). Fabricating the constructional steels. *Welding Design & Fab.,* 53–60.
4. Gifford, A. F. and Gorton, O. K., Total QA Makes for Successful Repair of Ammonia Converter—Fabrication and Reliability of Welded Process Plant. The Welding Institute, London, 1976.
5. Roberts, D. F. J. and Frazer, H. W., Report on Access for Welding. CEGB/Boilermakers' Collaborative Committee, July 1967.
6. Abrahams, C. J. (1970). A Designer's Guide to Inspection Problems. *Metal Construction,* **2**(9), 365–8.
7. Standard Data for Arc Welding. The Welding Institute, London, 1969, 129 pp.
8. The Procedure Handbook of Arc Welding. 12th edn. Lincoln Electric Company, Cleveland, Ohio, USA, 1973.
9. Johnson, L. Marvin, Quality Assurance Programme Evaluation. Stockton Doty Trade Press Inc., California, USA, 1970.
10. Clark, R. L. (Oct 1980). Investigate suppliers to maintain quality control. *Hydrocarbon Processing,* 171–86.
11. Gifford, A. F. and Orme, J. M., The Control of Welding Consumables for High Quality Products. The Welding Institute, London, 1974.
12. Clark, W. D. and Sutton, L. J. (1974). Avoidance of losses caused by confusion of materials in piping systems. *Welding & Metal Fabrication,* **42**(1), 21–5.
13. Welding Steel Without Hydrogen Cracking. The Welding Institute, London, 1973, 68 pp.
14. Colder, K., The TED Electrode Container. NEI Mechanical Engineering Ltd., Internal Report, October 1980.
15. Nicholson, S., Problems associated with the production and implementation of quality plans. *J. Brit. Nuclear Soc.,* **16**(3), 277–9.
16. Welding Approvals—A Testing Time for All. Sheffield & East Midlands Branches of the Welding Institute Seminar, April 1979.

17. Turnell, A. J. (1970). ASME welding procedure specifications and how to comply with them. *Weld Metal Fab.*, **38,** 355–9.
18. Boniszewski, T. and Pavely, D. A. (Nov. 1978). Timing at temperature when drying welding electrodes before use. *Metal Construction,* 530–1.
19. Nicholson, S. and Cresswell, H. (Sep. 1979). The calibration of welding equipment. *Metal Construction*, 422–5.
20. Control of Distortion in Welded Fabrications. The Welding Institute, London, 1976, 80 pp.
21. Anon, Construction of Littlebrook Power Station. *Metal Construction,* November 1980, 588–97.
22. Salter, G. R. and Gethin, J. W., An analysis of defects in pressure vessel main seams. *Pressure vessel standards—the impact of change.* The Welding Institute, London, 1972.
23. Gifford, A. F. and Blount, N. R., How Should Quality Costs be Measured in the Pressure Vessel Industry? Paper presented at the International Conference on Pressure Vessels, London, 1980.
24. Lochhead, J. C. and Speirs, A., The Effects of Heat Treatment on PV Steels. West of Scotland Iron and Steel Institute. **80,** 1972–3.
25. Smith, N. and Bagnall, B. I. (Feb 1968). The influence of sulphur on heat affected zone cracking of carbon manganese steel welds. *Brit. Welding J.*, 63–9.
26. Hart, P. M., Low Sulphur Levels in C–Mn Steels and Their Effect on HAZ Hardenability and Hydrogen Cracking. Trends in Steels and Consumables for Welding, The Welding Institute Conference, London, 1978.

4

Quality Control of Site Welding

B. S. Butler
British Steel Corporation, Middlesbrough, UK

INTRODUCTION

In recent years the importance of quality management systems has been recognised in most contracts involving welded fabrications and quality assurance has become a mandatory requirement in many cases, see Chapter 1.

The reasons for this development are largely historical and followed a number of expensive and potentially catastrophic failures [1, 2]. Elaborate systems have been developed with classical clearly defined lines of responsibility, authority and communication from design through to finished fabrication. Individual company QC departments are monitored by their own QA systems which in turn are audited by the client. On balance the net. effect of this activity has been a general improvement in the quality of fabrication on the shop floor. In the context of overall project control, it has meant an improvement in the quality of design and workmanship, a reduction in the frequency of expensive errors and ensures prompt delivery. The advantages have been obvious to the fabricator in terms of enhanced profitability and reputation and the user benefits from improved reliability.

However, management in general has been slow to recognise the benefits which established systems can offer to site construction. The advantages of QC are best illustrated after an incident when deficiencies in control are clearly the cause of the failure [3]. In other cases where QC is built into a site procedure [4] the advantages, though equally apparent, are less convincing.

Many factors militate against acceptance and they are, in the main, based on well established custom and practice. Site management is, by tradition, autonomous with complete authority over many functions which, in normal shop floor control systems, would be delegated. This position has evolved over the years because of the penalties for delay in the final construction stage, or in an on-stream plant down for repair, and the work programme tends to become more sacrosanct than ever. Whereas at the shop floor stage delays may eat into the site construction time, in the final stages of construction no such buffer exists. The pressure on site management to avoid delays from any source is therefore considerable and there is a reluctance to introduce any system which would complicate established procedures with a potential for causing delays. Usually greatest emphasis is placed on pure management skills which are obviously very important when dealing with an itinerant and quite often militant labour force which can present a considerable risk to progress. This emphasis is often detrimental to technical management and quality control, particularly in the area of welding and fabrication where the technology of design, materials and production has advanced so rapidly over recent years. Consequently there is an ever present risk that management will take technical decisions based on out-dated knowledge and experience which offers the least complicated and probably quicker solution without realising the possible consequences. This factor together with the present day reward systems which recognise the immediate benefits of a 'job on time' rather than the longer term effects of a 'job done properly' can put technical and quality management under considerable pressure.

MODIFICATION AND MAINTENANCE OF OPERATING PLANT

It should be remembered that site welding conditions apply equally to the modification and maintenance of operating plant. Process plant involving high temperature processes, particularly in the iron and steel industry where the product is fairly innocuous, apart from the obvious temperature hazard, erosion or accidental damage to the protective refractory structure can lead to severe damage to the fabricated containment vessel. This damage usually takes the form of a breakout of hot metal or severe distortion. Alternatively, particularly where thermal cycling has occurred, cracking may result. In either case the nature of the problem generally requires a major repair involving replacement of the damaged area rather than a local weld repair more usually associated with mechanical damage from, say, fatigue failure or brittle fracture. Repairs are required on an emergency basis and associated damage from the breakout or process debris and very poor access can inhibit the application of an ideal technology. The plant has to be returned to a safe working condition as quickly as possible with a view to a more permanent repair at a later, more convenient time. In cases such as this, which can be fraught with risk from almost every conceivable welding problem, there is an essential need for close technical control to ensure that the inevitable calculated risks are taken in a responsible manner to minimise the possibility of any cumulative effect arising from individual expedient measures.

In defence of site management it must be said that in many cases outside personnel involved in quality engineering, particularly in urgent situations, are not always blessed with the virtues of urgency. Obviously malpractice under any circumstances cannot be condoned but there *are* cases when technological guidance of practical experience is all that is required to assure quality and punitive imposition of standards can have a completely adverse effect.

CONDITIONS PECULIAR TO SITE WORK

The scale of site operations can vary considerably but characteristically the work is spread over a wide area and access for welding, supervision and inspection can be difficult and will be discussed in detail later. In the case of large industrial developments the geographical area covered and the size of the structures has to be recognised. For example, the British Steel Corporation's Redcar Development extended over a total area of approximately 1000 acres and included some 10 separate major plant construction developments together with ancillary plant. The blast furnace, the longest single structure on the site is some 17 m diameter at the base with an overall height of about 100 m and comprises plates up to 125 mm thick. There is little wonder that fabricators visiting such sites are often amazed at the relative insignificance of the fabrication which strained their shop fabricating capacity almost to the limit.

Where a number of major contractors and a multitude of associated sub-contractors are involved, supervision for quality can become difficult. Quality control is seen by many as a problem to be avoided rather than a system to be observed and some have yet to master the elementary techniques of quality control (see Chapter 3).

Bearing in mind these differences between shop and site environments the problems of quality management become obvious. Nevertheless, standards of construction equal to those achieved in shop conditions have still to be met on site.

THE CUSTOMER NEED FOR A CONTROL SYSTEM

Site operations can vary from a small local repair, involving a single welder, to a large multi-contractor project costing many millions of pounds. Quality and reliability is equally important in both cases, lapses in control can jeopardise the security of a whole plant. There are numerous examples of small, apparently insignificant faults arising from malpractice giving rise to major failures. For example a major crane gantry collapsed, resulting in the destruction of three overhead cranes due to a brittle fracture which initiated from a poorly repaired hole in the web of the beam (see Fig. 1). No reason could be given for

FIG. 1. Collapsed crane structures caused by a major fracture which can be seen in the main girder, emanating from a poorly repaired small hole in the girder web.

the hole in the beam, which probably arose from an error in manufacture and was not considered serious, but clearly was a major factor in this serious incident which, in addition to the complete loss of expensive capital plant, caused serious disruption to production. It is therefore of prime importance to establish an efficient system for dealing with quality aspects.

Most industrial enterprises which require welding services are sufficiently large to include a supervising engineer, or some engineering expertise. Therefore even the smallest, most insignificant contracts can be controlled by the customer who should have a responsibility for establishing his requirements in the contract and for making provision to ensure that they are satisfied. The actual character of the system which has to be established to provide this assurance depends on the size and nature of the contract.

LARGE AND SMALL CONTRACTS—SUB-CONTRACTORS' RESPONSIBILITY

In the case of small projects it may be that one contractor will complete all of the work himself. In such cases the responsible person in the customer's organisation will satisfy himself prior to work starting that the contractors' proposals for completing the work satisfy the requirements of the contract. In some cases it will be necessary for him to consult a separate specialist authority either within or outside his own organisation. At this stage agreement will be reached regarding the need for and extent of inspection prior to, during, and on completion of the job, guided by the contract specification. In many cases it may be that after the initial agreement on the *modus operandi,* the only control necessary will be inspection of the completed work, though this will be influenced by the confidence the engineer has in the contractor and his organisation.

Such small contracts involving close personal contact represent an example of the need for least independent supervision. If an individual is doing a job for himself, to his own specification as it were, there may be no need to apply any external control to ensure satisfaction. When he gives that job to another person it will be necessary to check that the work is satisfactory and the extent of such checks will vary depending on his confidence in the other person. Should the work be given to a third, fourth or fifth person the need for control becomes increasingly obvious.

This situation is typical of the conditions which exist in the contracting business and the longer the contract chain the greater the risk to quality. In larger contracts it is by no means unique for the chain to extend through 5 or 6 contractors. On larger scale multi-contractor operations such as major construction sites the importance of formalised control is paramount and it is a commonly held view that the customer should supervise the complete system. The main purpose of any system is to facilitate close liaison between the contractors concerned to ensure a clear understanding of quality requirements and that no discrepancies arise at contract interfaces in terms of design or standards of workmanship.

In order to avoid duplication of effort a quality assurance programme should be established early, under the supervision of an engineer/manager who should ideally report directly to the project director and have equal standing with the project manager in the ranking system as

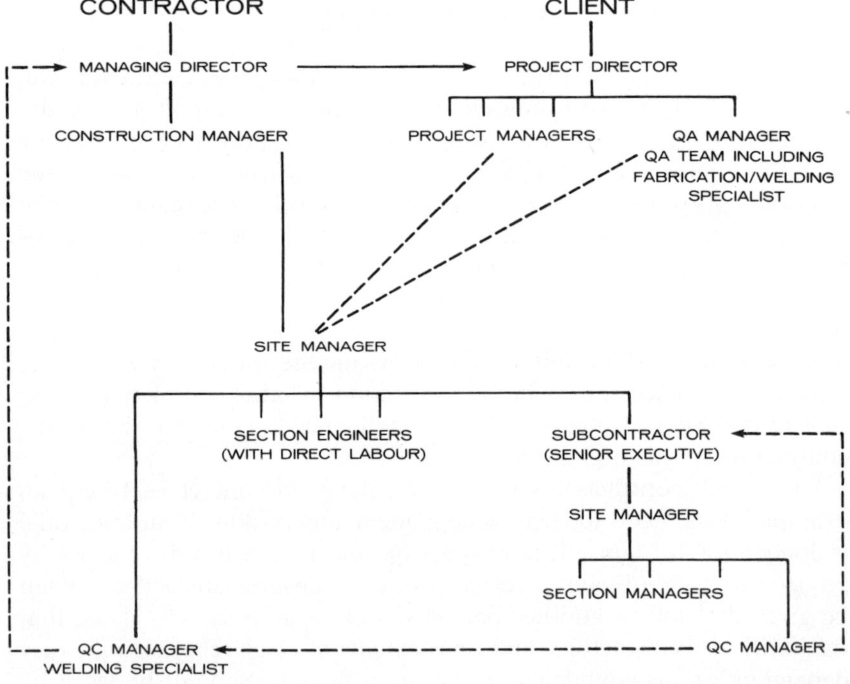

FIG. 2. Chart indicating the inter-relation between client and contractor in relation to quality requirements.

indicated in the idealised organisation shown in Fig. 2. This approach is slightly different from the proposal for contractors, which is also illustrated, where there are risks in disturbing the traditional autonomy of site management. The QA programme should co-ordinate all functions influencing quality across the organisation. Historical site problems causing major disruptions through strikes have encouraged close inter-contactor liaison on industrial relation problems particularly on recruitment and compliance with site agreements. Liaison through this system can establish a pattern for personnel control in the welding and associated trades which will benefit the activity as construction proceeds. For example, in addition to ensuring that all personnel are members of the appropriate trade union and have no record of disruptive activities, it would be a simple matter to ensure that they have the relevant practical ability. It is not unknown for even quite reputable contractors, faced with a desperate situation, to recruit 'skilled' labour without any check on ability! For high quality work welders will be screened in the mandatory qualification test procedures but where such tests are not required the consequences of inferior workmanship can be at best delaying and at worst potentially catastrophic. Therefore, all welders who are employed on site should be registered with the central site organisation and through that the quality assurance system where their credentials can be checked and the need for further surveillance considered.

DEVELOPMENT OF THE ORGANISATION

With the principle of quality assurance established there is a need to develop the associated support organisation. The size of the organisation will obviously depend on the size and complexity of the contract to be supervised.

Where a plant is in operation and repair work or modifications are to be undertaken, the established technical functions are usually capable of dealing with quality aspects and the works engineer or technical manager may take on the temporary responsibility for quality. The nature of the ultimate product may have some bearing on the detailed staffing of the organisation. For example, in a metallurgical industry technical personnel may be more conversant with welding and fabrication problems than, say, their counterparts in non-metallurgical industries, since it is in the main a metallurgically based technology.

However, the principles of QA should be familiar to all engineers and technologists. It may not be essential for QA personnel to have detailed knowledge of the relevant fabrication processes but in order to retain credibility, without which they will lose the respect of the contractors and thereby adversely affect quality, it is essential that they have direct and immediate access to metallurgical/welding expertise when required.

Generally a good engineering background and a familiarity with engineering standards, control and assurance systems in so far as they affect welding and fabrication are suitable qualifications for QA personnel. The manning levels will depend on the size of the project and the facilities available from the contractor and his sub-contractors. The site conditions and the nature of the work should be taken into account bearing in mind the problems already discussed in relation to shop construction. The need for clerical staff to cope with the documentation which can often overburden technical staff should be recognised from the start. In 'greenfield' developments where there is no resident organisation a separate clearly identified function should be established in line with that detailed in Fig. 2.

THE SCOPE OF SITE QC—CONTRACTOR RESPONSIBILITY

Having established a site QA organisation it is now necessary to establish a control policy. The classical controls apply to both the designer and erector alike. The conceptual design, general arrangement and detailed design should be audited to ensure that they meet the customers' requirements from an engineering standpoint, can in fact be made and will function. This work is often outside the scope of the QA organisation who should be responsible for ensuring that the work is done effectively, taking whatever steps are necessary to ensure the competence of the assessors.

It should be the prime responsibility of the site QC organisation to ensure that the construction stage can be and is satisfactorily completed by the contractor starting with critical resource assessment through labour recruitment, welding approval and production capability surveillance to final acceptance. Particular attention must be paid to capability and resources. Site operations are, in general, serviced by reputable contracting organisations with recognised ability and skills.

However, inevitably from time to time these contractors approach and pass their boundary of competence and expertise and it is in this area that they, together with other less reputable companies, represent the greatest risk to a project. It is therefore vitally important that claims of previous experience supporting their proposals are thoroughly checked in the course of the normal resource assessment. This checking will vary depending on the size and nature of the contract but must involve very close technical questioning to ensure that any claimed experience is compatible with the current job requirements. The type of questions to be asked in order to make a sound judgement of competence are given below. This information may be obtained in a written questionnaire but any apparent deficiencies which are revealed must be pursued in verbal discussion.

(1) Have you completed a job of similar size and complexity? Supply detailed drawings and contract specifications. State location of site and climatic conditions.
(2) What materials were involved? State types, thicknesses and any special quality requirements such as through thickness, ductility and controlled carbon equivalent.
(3) What were the fabrication techniques used for erection? Supply details of fabrication procedures.
(4) What were the welding techniques used? Supply details of welding procedures including non-destructive test requirements.
(5) Are there any design changes in the current contract which affect welding details?
(6) From your earlier experience what special arrangements, i.e. labour recruitment, procedure and welder qualification and supervision, do you consider necessary to guarantee quality?

Site contractors are often criticised for the apparently over confident approach to their work. In many cases it is not important and indeed without this confident, if even sometimes arrogant attitude there are many aspects of site construction work, seen more recently in the offshore developments, which could not be contemplated. It is however important to be sure that the experience of the contractor has a sound technological base and that the current contract is within his technical capability.

Precautions should not be restricted to small contractors, of whom there is a tendency to be automatically suspicious. Large international contractors can be very impressive with articulate accounts of previous

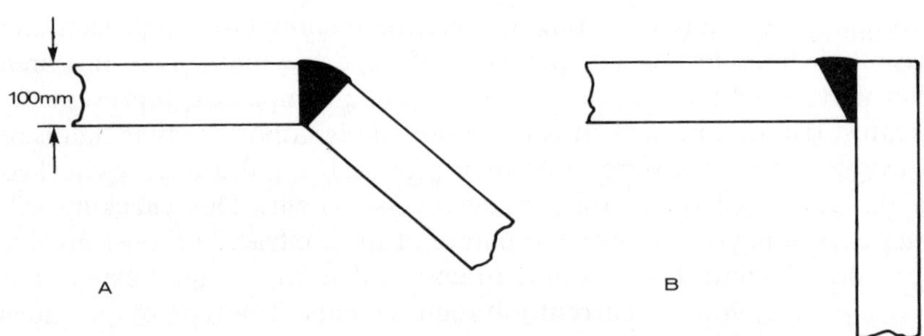

FIG. 3. Influence of a simple design change on potential welding problems. A, detail not susceptible to lamellar tearing; B, detail highly susceptible to lamellar tearing.

work carried out all over the world. Experience has shown that in many cases they cannot recognise the strains which an increase in size or a change in materials will place on their established technology. As an example in recent years, lamellar tearing in fabricated steelwork has been a major problem and it is sometimes extremely difficult to persuade major international contractors that there are potential problems with proposed designs. It may be claimed by the contractor that he has constructed a number of plants similar to the one in question without problems and it may not be until the details of the earlier work are disclosed that the reasons become obvious. Such a case is illustrated in Fig. 3 which shows the two welding details used for constructing the end of a ball mill. In the earlier design (A) on which a contractor based his confidence, the joint was immune from the risk of lamellar tearing whereas the detail proposed (B) is an obvious candidate. The costs involved in protecting against the problem were small compared with the potential costs of repair and the inevitable delay in completion of the work.

The contractor's organisation must have a QC system capable of dealing with the demands of the contract to the satisfaction of the customer's QA function. If it is not suitable and cannot be improved, it is essential that the customer's site QA system is capable of, and has the authority to deal with QC from receipt, and storage of materials and welding approval through to final acceptance of the finished job.

Goods Receipt, Storage and Release

The contractor's site management team is usually responsible for the quality of the finished construction irrespective of the source of supply of component parts. It is therefore essential to ensure that all materials and equipment delivered to site meet the specification requirements and fabricated equipment which is usually more vulnerable to damage should be thoroughly inspected before acceptance. This procedure provides a 'safety net' for faults left undetected at the shop inspection stage and covers any damage which may have occurred in transit. Quite often equipment delivered to site has to be stored for unpredictably long periods, particularly when delays to progress on site occur. Errors discovered at an early stage can often be corrected with minimum effect on the overall programme. The possibility of prolonged storage should also be anticipated in connection with weather protection. Sensitive items such as welding consumables and equipment should be stored under cover and where this is not strictly necessary or practicable, suitable precautions should be taken to minimise the risk of damage. For example, structural fabrications designed for external use should include drain holes to avoid entrapment of water in-service. If the fabrication is placed on its side during storage, the drain holes may be completely ineffective resulting in quite severe deterioration by corrosion. There are many examples of enclosed fabrications containing sensitive electrical equipment, e.g. overhead cranes. Although the general standard of welding may be good the unit may not be absolutely waterproof and deterioration of the electrical contents may occur, particularly where severe temperature fluctuations result in condensation damage. The message in these examples is to store equipment efficiently and inspect regularly to ensure that protection is maintained.

Storage and Issue of Welding Consumables—Electrodes, Fluxes and Wires

The care of fabrication consumables, particularly those for welding, should be equivalent to that available for shop manufacture unless the production procedures to be used can tolerate a lower standard. It would be folly to produce a welding procedure for site which is more demanding than shop practice. Generally circumstances dictate that the procedures are the same. The extent of control required will be influenced by the construction. In critical cases incoming materials may have to be bonded pending release checks for conformation to specification but generally, for structural welding, this precaution is not necessary and quality assurance by the consumables supplier is adequate. The storage area must be enclosed, dry and free from violent temperature fluctuations. In most cases, provided these requirements

FIG. 4. Brittle fracture in end plate of low pressure heat exchanger initiating from arc strikes on plate surface.

are followed and additional baking or drying of electrodes and fluxes carried out in accordance with the manufacturer's recommendations, problems associated with hydrogen contamination can be avoided.

The cause of cracking problems in the weld and heat affected zone (HAZ), such as that involving a heat exchanger end plate, Fig. 4, where hydrogen induced HAZ cracks were associated with local arc strikes, can in many cases be traced to the lack of attention given to the care of electrodes during storage. In this particular instance the electrodes were stored in an unheated corrugated sheet hut with condensation dripping down the walls. Such abuse of welding consumables is not uncommon but must be avoided. It is frequently proposed that the foreman's office be used as an electrode store. This can be equally unacceptable, particularly when the levels of relative humidity and temperature are not controlled. Equally important are electrode baking/drying ovens and cabinets. They should allow free air circulation and have efficient temperature control. Damp electrodes cooked or 'broiled' in an airtight container will not dry no matter for how long they are treated.

On a more general note, a neat well-managed stock using materials on a rotational basis reduces the risks from deterioration or damage and ensures against waste from obsolescence when old stock has to be cleared out and scrapped. Such 'good housekeeping' also reduces the risk of electrode mixing. In many cases the consequences of this latter risk are negligible if there is a limited range of similar materials being used on site, but in cases where specific electrode properties are required, an electrode mix-up can be very serious indeed. An obvious example would be mixing of rutile for basic coated electrodes which could lead to hydrogen assisted cracking problems. There are many other equally serious examples where the weld metal mechanical or physical properties required for the service conditions may not be achieved resulting in premature failure of the plant with all the associated consequences.

WELDER SELECTION AND APPROVAL

Control of welders on site is vitally important to quality. The psychological effect of active surveillance is often sufficient to ensure the required quality and will avoid the risk of bad practices. Its value should never be underrated. However, deployment of a suitable workforce is the first consideration. On large construction sites, in times of industrial boom or in emergency repair situations, competition for labour can be fierce and there is a temptation to recruit without giving thought to competence. In the case of higher quality fabrication work demanding detailed welder and procedure qualification, poor welders will generally be screened out. However, for work not covered by these requirements, there is a real risk of attracting the lower levels of capability and competence. This situation can be forestalled by ensuring that each welder is examined for practical skill before being employed. This examination may simply involve observation of his welding technique by a skilled assessor or take the form of a test to a recognised code, e.g. BS 4872 Approval Testing of Welders when Welding Procedure Approval is not Required. Production work should be closely examined until the supervisor is absolutely confident of the welder's ability.

WELD PROCEDURE DOCUMENTATION

All procedures for welding must be clearly defined. For simple structural welding the requirements should be covered in the engineering specification. For example, simple single run fillet welds in low strength carbon and carbon–manganese steels can be fully specified by notation on the drawing in accordance with a national code such as BS 499 Welding Terms and Symbols. More critical welding on important items of plant or on thicker, less weldable materials will require precise welding details to be recorded in a welding procedure document which should be agreed between the parties concerned well before commencement of fabrication and may have formed part of the pre-contract documentation.

For pressurised systems and some process plant the requirements for documentation are set down in nationally applicable standards. However, a considerable amount of structural fabrications, together with some very heavy process plant and mechanical handling equipment, is not covered by existing national standards. This deficiency led the British Steel Corporation to develop its own in-house standard [5] which was discussed at the Welding Institute Autumn Conference in 1976 [6].

The standard classifies equipment under three categories in descending order of quality. For the most critical conditions a detailed welding procedure proposal document (Fig. 5) is required for approval before the contract is placed. The proposal should include all details of the welding procedure including heat treatment and inspection. On the basis of this document details of procedure qualification tests are decided which should be completed well in advance of production welding to allow for any re-testing required.

WELD PROCEDURE TESTS

The scope of a site procedure test must cover the work in hand and should be witnessed by the site welding engineer who should be able to

CORPORATE ENGINEERING STANDARD—WELDING OF STEEL FABRICATIONS

WELDING PROCEDURE PROPOSALS—CATEGORY 1 WELDS
This document on completion shall form part of the contract.

CONTRACTOR:- WELDING PROCEDURE FOR:-
FABRICATOR:- DRAWING REFERENCE:-
PREPARED BY:- CHECKED BY:- APPROVED BY:- DATE:-

WELD PREPARATION—INCLUDING TOLERANCE ON FIT-ACTIONS ON DEVIATIONS TO BE NOTED IN 'SPECIAL PRECAUTIONS'.	PASS LOCATION AND SEQUENCE

JOINT LOCATION		
MATERIAL		
METHOD OF PREP.		
WELDING PROCESS	1st Side	
	2nd Side	
PREHEAT	Temp °C	
	Method	
	Retention	

Pass	Electrode Size	Electrode Code No.	Weld Position	Amps	Volts	Speed

WELD MATERIALS	1st Side	2nd Side
Filler Metal		
Power Supply		
Polarity		
Flux or Gas Shield		
BACK GOUGE METHOD		
BACK GOUGE INSPECTION		
HEAT TREATMENT	Stress Relief	Normalising
Heat Rate °C/h		
Soaking Temp °C		
Soaking Time h		
Cooling Rate °C/h		
Withdrawal Temp °C		
WELD FINISH		

SPECIAL PRECAUTIONS AND NOTES

N.D.T.	

FIG. 5. Procedure proposal document (from BSC CES 22, reference 5).

recognise potential problems associated with a site environment. The actual procedure approval test requirements may exceed those specified in the normal national application codes. There is no value in completing a test weld with a shop based machine when the site welds are to be made with a modified version of the equipment. In some cases, particularly where there is no previous experience the procedure test will need to take the form of a full scale demonstration. Girth welding of larger cylindrical components is a case where a simple horizontal/vertical (e.g. ASME IX 2G detail) butt weld produced under shop conditions would not demonstrate the suitability of the equipment on site. In order to generate the necessary confidence a large scale mock-up should be used to demonstrate both the welding and mechanical characteristics of the equipment. It should also be recognised that practice with the equipment prior to commencement of production welding will reduce the on-the-job learning and benefit production. There are many instances of introducing sophisticated procedures into a site environment without recognising this fact with disastrous consequences. Electroslag welding on site, for example, is not very common and on some thicker sections a sophisticated modification of the standard equipment is required to produce a satisfactory weld. Such modifications may appear simple from a mechanical viewpoint but it must be recognised that the operation can become very complicated and any incompetence can lead to very serious problems.

Having stressed the need to formalise acceptance of procedures there are emergency situations particularly in repair and maintenance, where time will not be available to follow this idealised route. In such cases a thorough technical assessment of any proposals must be made by a competent person, before any work is allowed to proceed and testing required to prove the procedure put in hand immediately. Where a very urgent repair is required and a quick short term solution is a viable proposition this must only be undertaken on the authority of a competent person and a long term repair compatible with the design and service requirements must be undertaken as soon as practically possible.

SUPERVISION FOR QUALITY

The written procedure proposal (Fig. 5) which should be a contractural document, also forms the basis for quality surveillance against which workmanship is judged. Control of paper documentation tends to lapse more easily on site than in the workshop. The type of personnel attracted by site environment are usually more independent and self-assured than some of their shop floor counterparts. It is difficult and often unreasonable to insist that each welder or operative should be in possession of a written procedure but at least his immediate superior should have one, and be fully conversant with its requirements in order to instruct the welder accordingly.

This now raises the question of supervision of fabrication and welding. Site operations tend to be controlled by engineers in charge of sections of the project with little, if any, training in welding technology. Their knowledge is based on previous contracts and, in general, they only become familiar with specialist subjects such as welding when they have given rise to trouble in the past and they are able to recount the actions taken in particular circumstances, often without an in-depth appreciation of the technology involved. Their ability, therefore, to foresee problems and take evasive action is restricted and it is vitally important that they have the services of a welding technologist. The source of this facility will depend on the type of contract. Large contractors should carry a welding engineer who can supervise the work of a number of section engineers, perhaps even on different sites. Direct control of the labour force should be through a fabrication/welding foreman who will have a functional responsibility to the welding engineer for technical matters. All of the proposals for welding and the associated technology should be fed through this engineer who should be responsible for and have absolute authority in all matters relating to welding technology. His actual lines of responsibility can be difficult to define. He will have a responsibility to the site management for day-to-day administration but in cases of disagreement with regard to policy in technical matters affecting quality he must have a facility for recourse to a higher authority, possibly through his, the contractor's, QC system as indicated in the organisation chart in Fig. 2.

INSPECTION AND TESTING

Aspects of inspection in so far as they are affected by site construction will be discussed later. Weld testing on site should meet the same standards as shop construction and operatives should be qualified to a standard appropriate for the tests to be carried out, e.g. The Welding

Institute CSWIP (Certification Scheme Weldment Inspection Personnel). Usually because of the congested situations on site and problems of isolating areas for radiographic inspection, ultrasonic testing is preferred for the detection of internal defects. Equipment and personnel are now available for a high degree of confidence to be achieved with this technique.

STRESS RELIEF OF SITE WORK

Stress relief on site usually involves local heat treatment. However in exceptional circumstances where large components are either erected from small plates or transported in a number of prefabricated sections for final assembly on site, complete stress relief of the whole structure may be necessary. This can be achieved by either constructing a light weight sectional furnace to hold the structure or, in the case of vessels,

FIG. 6. Site welding of the closing seam in the trunnion ring for a large basic oxygen steel making furnace. The figure shows the seam being prepared for preheat and subsequent stress relief.

cladding the outside of the vessel with mineral wool insulation and heating from the inside either electrically or by gas.

Local heat treatment is carried out in a similar way usually with heaters attached close to the weld seam. Heating can be by gas, electricity or by specially prepared exothermic pads. In the later case the heat treatment is set by the volume and properties of the exothermic material and control during the operation is not possible. It is most usually applied to pipe work where many similar welds have to be treated. Local heat treatment should be in accordance with the requirements of an appropriate standard, e.g. BS 5500 1976 Specification of Unfired Fusion Welded Pressure Vessels which details the procedure necessary to ensure stress relief. Figure 6 shows the closing seam in the trunnion ring for a large basic oxygen steel making furnace being prepared for site welding. The electrically heated elements were used to heat the joint for welding and also the subsequent stress relief. Local weather protection is being erected over the joint area.

Vibratory stress relief has been shown to be beneficial in certain cases particularly for machining stability. However care should be exercised where components are subject to fluctuating stresses in service since vibratory stress relief can reduce the fatigue life.

HAZARDS TO WELDING QUALITY ON SITE

Many of the welding techniques and procedures used on site are similar to those widely used in shop manufacture. The main differences are in the conditions under which they operate. Repair situations in heavy industry can involve very dirty, cramped and exposed conditions and quite often very little can be done to improve the situation. On new plant constructions the location and accessibility of site welds should be considered at the design stage. However occasionally due to an oversight or force of circumstances access for site welding is poor and together with either construction or process debris this can have a significant effect on weld quality.

Dirty Conditions

Cleanliness is a factor in 'good housekeeping' already raised under electrode storage and clearly every attempt should be made to ensure that reasonable conditions are maintained in the welding areas to produce the best psychological conditions for a welder to work under.

This advisory note only applies to the general area. Local to the weld cleanliness must be mandatory. When components are contaminated with grease or oil they must be cleaned by a recognised system and in critical cases to a procedure approved before work starts. Quite often meticulous care is taken in removing visible contamination without realising that the heat from the welding process will melt other grease and thick oil, distant from the welding operation which may drip or seep into the weld area. In some cases this may be unavoidable and the consequences must be accommodated in the specified welding procedure and fully recognised before the work starts. There are no grounds for accepting defects which have arisen after all reasonable precautions have been taken. It is pointless using a procedure based on 'low hydrogen' consumables in such conditions. Better to recognise the problem and reduce the hydrogen risk by other means such as increased preheat and controlled post-weld heat treatment.

Accessibility

Whether accessibility is a problem arising from poor design in a new construction or failure of plant in service, the welders comfort and confidence is of prime importance. Comfort can be subjective and complete satisfaction may be impossible. Quality depends to a large extent on the skill and confidence of the operator but over confidence can be as dangerous as uncertainty. If a weld is difficult to make then it is certain to be difficult to repair and the quality achievable is significantly reduced with each successive repair. Therefore caution should be exercised when allowing welders to estimate their own ability, particularly in a site environment where the itinerant workforce tends to be very self assured by nature. Practice on a mock up simulating the actual conditions is recommended and should demonstrate the welders capability before the production welds are attempted. It is worthwhile to remember the maxim 'If a weld is difficult to make it will be difficult to inspect'—and quality assurance may have to be based on confidence in the welders skill and competence. Figure 7 illustrates a typical situation of poor accessibility in boiler manufacture where a very high standard of quality is demanded notwithstanding the practical problems.

Exposed Conditions

In common with most human activities where the final operation in a sequence of events is given the least consideration, welding often

FIG. 7. Illustration of restricted access for welding typical of site situation.

suffers from cumulative errors in planning. Quite often very important components which should be welded in controlled conditions are required to be welded in the open air. In ideal weather conditions for welding this may be tolerable but it is certainly unacceptable in a more changeable climate. It is similarly unacceptable to have roof cover without adequate side wall protection. Quite often equipment is installed in buildings which are partially complete and from which the side cladding has to be left to allow access for additional heavy equipment. The protection necessary in such cases will depend on the nature and size of the work but in critical situations or where a considerable amount of work has to be done, complete protection with tarpaulins or polythene sheets is essential. Figure 8 shows the local protection

Fig. 8. View of a large blast furnace under construction showing the temporary protection for the welding operation.

erected around a blast furnace shell during welding of the main seams and Fig. 9 illustrates the complete encapsulation of a 40 ft diameter sphere in order to ensure controlled conditions.

Security of Equipment

A serious problem on most construction sites is security of equipment. This should be recognised when planning or executing inspection schedules. Preheating torches, welding cables, earthing clamps, lights and other ancillary equipment are often mislaid and not replaced. This leads to risks being taken with the approved procedure and the appropriate workmanship standards. For example poor earth connections can lead to arc strikes and unstable welding conditions and the effects of inadequate illumination are obvious. Absence of suitable heating torches can induce a welder to dodge the preheat requirements. Similarly welding equipment, power supply and other services should be maintained at a level which can achieve acceptable quality. Frayed or damaged welding cables, damaged or inefficient welding

FIG. 9. Forty ft diameter sphere completely encapsulated by temporary screening to ensure controlled conditions for site welding.

FIG. 10. Photomicrograph of a repair weld showing severe carbon contamination from inefficient air arc gouging.

equipment or overloaded power supply can all influence the weld quality directly or indirectly through the frustrations caused to operatives. An extreme example of the effects of inadequate air supply in air arc gouging is shown in Fig. 10. In this case the molten metal pool created by the arc has not been ejected by the air stream and the carburised material, on subsequent solidification, has formed pools of hard cementite associated with martensite resulting in severe cracking problems.

An efficient maintenance force preferably based on prevention rather than cure is therefore as important to quality as it is to productivity and this is one of the few instances where the requirements of production and quality function are compatible.

REFERENCES

1. Report on Brittle Fracture of a High Pressure Boiler Drum at Cochenzie Power Station. South of Scotland Electricity Board Report, January 1967.

2. BWRA Bulletin (June 1966) **7**(6), 149–78.
3. Clark, W. D. and Sutton, L. J. (1974). *Weld. Met. Fab.*, **42**(1), 21–5.
4. Gifford, A. F. and Gorton, O. K., The Welding Institute Conference, November 1976, Paper 22, 115–27.
5. British Steel Corporation, CES 22, Welding of Steel Fabrications.
6. Garland, J. G. and Butler, B. S., Welding Institute Conference, November 1976, Paper 21, 103–13.

5

Defects in Welds—Their Prevention and Their Significance

J. H. ROGERSON
Cranfield Institute of Technology, Bedford, UK

INTRODUCTION

Welding is a complex technology and in any welding process (including mechanised processes) the quality of the weld is a function of the interaction of a large number of variables, not all of which are controlled to the extent that is desirable. The difficulty of completely controlling the welding process means that most welds will contain some defects even if no 'errors' are made in selecting the materials, joint design or welding procedure. Also the continuous improvement in non-destructive testing methods means that such defects as are present are increasingly likely to be detected.

This situation leads, first, to a need to understand the cause of weld defects and how to prevent them occurring as far as is possible. Obviously any fabricator's aim must be to prevent the production of defective welds. However, the difficulty and the cost of consistently producing welds without *any* defects is such that we must come to terms with the fact that welded structures produced at an economic cost may contain a proportion of weld defects. We need to know therefore the significance of the different weld defects in terms of weld performance so that we can define a safe and realistic tolerance limit for defects for each class of welded structure. It is, of course, a key factor in the quality assurance of any component or product that the specification of the quality standard be appropriate.

DEFECT TYPES AND THEIR EXPECTED FREQUENCY

The International Institute of Welding has proposed [1, 2] a comprehensive classification of defect types. In this classification six main groups of defect are identified:

(1) Cracks;
(2) Cavities (porosity and shrinkage cavities);
(3) Solid inclusions (slag, oxide);
(4) Lack of fusion and lack of penetration;
(5) Imperfect shape;
(6) Miscellaneous (e.g. spatter, arc strikes, grinding marks).

It is necessary to look at this classification in terms of the main causes of the defects, i.e. whether they are 'technological' defects or 'workmanship' defects. 'Technological' defects are defined as those resulting from a major inconsistency in the welding operation, such as a wrong electrode, incorrect heat treatment, inappropriate joint design, whereas 'workmanship' defects are those which arise from the inherent variability of the welding process or a chance error by an operator. The majority of the cracks and lack of fusion and lack of penetration defects can be considered as technological defects whereas the majority of cavities, solid inclusions, shape and miscellaneous defects can be considered as workmanship defects.

The amount of quantitative information on the number, size, type and distribution of weld defects in different classes of welded structure is not very extensive. The reliability of some of this information is also open to question because of the unknown reliability of some inspection methods (in particular ultrasonic inspection) and the unknown reliability of defect reporting methods.

Two pieces of work which are frequently quoted are that due to Salter and Gethin [3] who analysed defect lengths in radiographs of main seams of pressure vessels and that due to Kihara *et al.* [4] who assessed the proportion of radiographs which indicated 'unacceptable' defects for a range of structures. Defect 'rates' for pressure vessel welding were of the order of 3 % whilst for lower quality welding defect 'rates' of up to 20 % or more were found. In all cases, though, the majority (usually an overwhelming majority) were minor workmanship defects. This general pattern has been confirmed by later work [5, 6] for a wide range of structures and Table 1 derived from the data in references 5 and 7 indicates this.

The distribution of the workmanship defects can be considered to be random [5, 7] and this is not surprising as such defects arise from chance (random) locally occurring variations in the welding conditions. The inherent variability of the arc, occasional errors in electrode

TABLE 1
COMPARATIVE AVERAGE DEFECT RATES (ALL INDICATIONS ≥2 mm RECORDED)

Structure type	*Average number of defects per metre of weld*
Aluminium pressure vessel (MIG welded)	1·6
Steel site welded tankage (MMA welded)	0·7–4·9
Steel site welded tankage (Sub arc welded)	3·4
Ships hulls (MMA welded)	4·9
Low alloy steel pressure vessel (MMA welded)	0·7–1·0
Low alloy steel pressure vessel (Sub arc welded)	0·3–0·5
Nodes in tubular offshore platform jacket (MMA welded)	5·9[a]

[a] Derived from ultrasonic test data, all other data derived from radiographs.

manipulation, slag removal or fit-up, and the consistency of operation of the welding equipment are all events which occur randomly and which can result in minor workmanship defects. The major, technological, defects are defects resulting from significant errors (incorrect electrode, incorrect procedure, for example) and these will not be random events. Therefore there is no useful distribution function which we can use to describe the occurrence of such major defects.

The size distribution of defects is of importance because it is relevant to the detectability of defects by non-destructive testing. Clearly we would expect small defects to predominate and large defects to be much less frequent. This is, of course, the case but it is difficult to describe size distributions in a quantitative manner. Various sources of data exist, notably for PWR vessels [8] and for welds in offshore structures [9]. Exponential [9, 10] or Weibull [9] distributions have been variously assigned to this data. The parameters of distribution, whichever distribution is used, will obviously vary with the quality levels of the fabrication although in the case of offshore structure welds a surprising degree of similarity has been found in the size distribution of defects in different structures.

CAUSES OF WELD DEFECTS AND THEIR PREVENTION

Technological Defects

A major type of technological defect is cracking which can be from various causes but all, essentially, a function of microstructure and stress:

(1) Hot (or supersolidus) cracking—possible in all alloy systems.
(2) Reheat cracking—almost exclusively restricted to creep resistant steels.
(3) Hydrogen induced cold cracking—ferritic steels only.
(4) Chevron cracking—high strength weld metals in ferritic steels only.
(5) Lamellar tearing—in principle possible in any material but in practice restricted to structural and pressure vessel ferritic steels.

Hot (or Solidification) Cracking

This is intergranular cracking (see Fig. 1) which occurs during or just after solidification and is normally found in the weld metal although a similar form of defect can occur in the heat affected zone immediately adjacent to the weld. During solidification weld metals pass through a temperature range in which the metal has a very low ductility and so cannot easily accommodate the localised strain resulting from the differential expansion and contraction of the weldment due to phase changes and the restraining imposed by the inherent properties of the partially solidified metal, i.e. interlocking of dendrites in the last stages of solidification [11, 12] or the presence of low melting point liquid films such as iron sulphides or iron–iron phosphide eutectics in ferritic steels [13].

Further subdivisions of the phenomenon have been proposed [14] but in all cases the cracking is a function of composition and stress so that the prevention of this defect type relies primarily on compositional control and, to a lesser extent, on the control of joint detail and procedure to reduce welding stresses. The main principle therefore is to select a weld metal composition which has a minimum freezing range so that the time the weld metal is in the low ductility region is at a minimum.

In the case of ferritic steels we find that sulphur, phosphorus, boron and niobium are the most harmful elements in terms of their effect on the solidification range. Therefore it is desirable to reduce the levels of

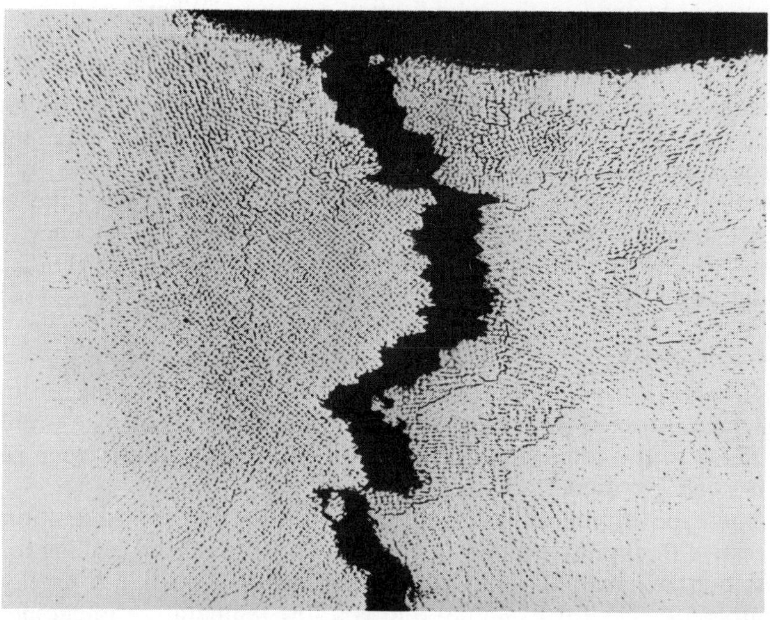

FIG. 1. Intergranular nature of a hot crack in a carbon–manganese steel (×50).

such elements as much as possible. A complicating factor is that these elements are less soluble in austenite than in ferrite so that elements which promote austenite formation rather than ferrite formation on solidification via the peritectic reaction increase their severity. For example, elements such as carbon and nickel can be considered as undesirable from this point of view.

In practice the prevention of hot cracking by compositional control in ferritic steels relies on the control of sulphur and phosphorus to very low levels, minimising carbon contents and restricting nickel contents to 1·0 % or less. Manganese which is normally present in ferritic steels helps to reduce the effect of sulphur by forming high melting point complex sulphides which effectively reduce the freezing range. Clearly the greater the manganese content the greater the tolerance for sulphur.

In austenitic steels prevention of hot cracking is frequently obtained by adjusting the composition to ensure that a minimum of 5–10 %

ferrite is obtained in the solidification structure. Where other considerations (e.g. corrosion control) forbid this approach then prevention must be by procedure and joint design control (i.e. limiting the stress). In non-ferrous metals it is frequently the major alloying additions which determine the freezing range (e.g. magnesium in aluminium–magnesium alloys) and so hot shortness control is achieved by the selection of non-matching consumables to significantly alter the weld metal composition for alloys which have a hot shortness tendency. The approach is typified by the analysis of the situation for aluminium alloys given by Young [15].

Reheat Cracking

This is a very serious (and difficult to rectify) cracking problem which is restricted to low and high alloy steels and is mostly a problem confined to the pressure vessel industry particularly where creep resistant steels are used.

One type of reheat cracking [16] is caused by the generation of excessive thermal stress during post-weld heat treatment leading to the initiation of cracking from preexisting defects (small hot cracks or hydrogen cracks for example) and is a low temperature phenomenon (~300°C) of thick section vessels in low alloy steels. The prevention of this is by control of heating rates and temperature distributions and by the avoidance, as far as possible, of stress concentrations.

The second, and more intractable, form of reheat cracking [16] occurs at higher temperatures (temperatures within the creep range) where intercrystalline cracking in the coarse grained heat affected zone (and occasionally in the weld metal) results from insufficient creep ductility. This cracking occurs either during post-weld heat treatment or during high temperature service and arises because carbide precipitation and impurity segregation strengthens the matrix to such an extent that creep strain is accommodated by grain boundary cracking. This defect can be prevented by correct selection of material composition and heat treatment. In ferritic steels the carbide formers molybdenum, vanadium and chromium are the most detrimental elements together with the impurity elements tin, antimony, arsenic and phosphorus. In austenitic steels niobium is the carbide forming element which causes the most problems and the use of molybdenum bearing steels instead is a prevention method.

Welding procedures which avoid the production of an excessively coarse grained heat affected zone (i.e. low heat input procedures) are also helpful in preventing this type of cracking as is a reduction in local stress levels (e.g. grinding of weld toes to reduce stress concentrations) and, in extreme cases, the selection of weld metals with a low hot strength.

There is a third type of reheat cracking similar to the creep cracking phenomenon which is the underclad cracking sometimes found in low alloy steels for nuclear vessels when clad with austenite steel. This very specific problem has been described in considerable detail elsewhere [17].

Hydrogen Induced Cold Cracking

This transgranular cracking phenomenon is associated with the heat affected zones and occasionally weld metals of ferritic steels. It occurs at low temperatures (<150°C) and sometimes only appears some hours after welding. This cracking is of very characteristic appearance (Fig. 2) and usually originates at a weld toe. It is caused by the diffusion of

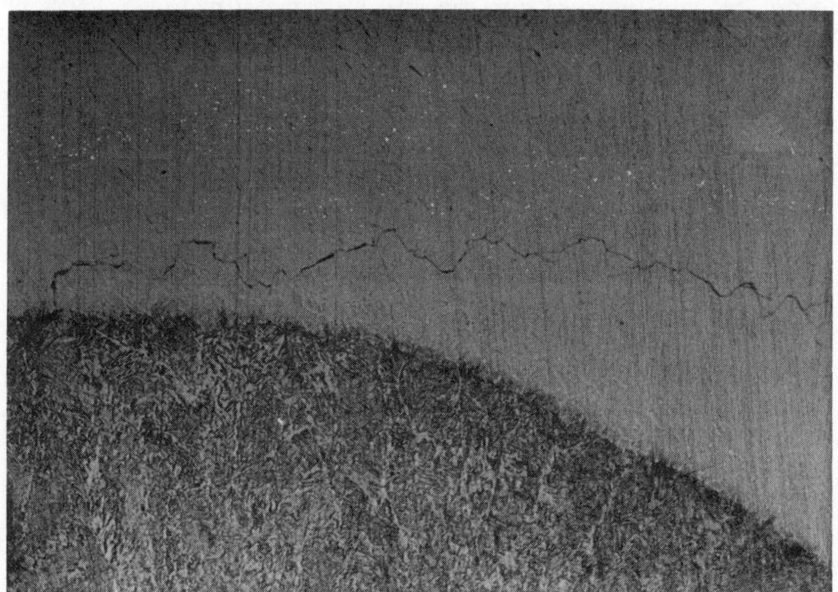

FIG. 2. Hydrogen induced cold crack in the martensitic HAZ of a steel of high carbon equivalent where insufficient preheat has been used.

hydrogen from the weld pool into martensitic structures. This embrittles the martensite such that local strains resulting from excessive external restraint or differential expansion and contraction will cause cracking. Therefore, for this type of cracking to occur it is necessary for there to be a martensitic structure, a sufficient amount of hydrogen present to cause embrittlement and a sufficient stress to cause cracking of the embrittled structure.

This is a well understood welding phenomenon and the prevention and control of it is obtained by the control of microstructure, hydrogen level and/or stress level. For the majority of ferritic steels practical guidelines for their prevention and control are given in the form of tables and nomograms in Appendix E to BS 5135 : 1974 Specification for Metal Arc Welding of Carbon and Carbon–Manganese Steel.

Control of hydrogen level: This is accomplished by the use of 'low hydrogen' consumables as the hydrogen in the weld pool mostly comes from moisture associated with fluxes. Manual metal arc electrodes and submerged arc fluxes are available which, if properly baked and stored to prevent moisture pick up, can give weld metal hydrogen levels of less than 15 ml/100 g whereas TIG and MIG processes being fluxless can give even lower levels (5 ml/100 g or less). A more extreme measure is the use of austenitic electrodes (the diffusion rate of hydrogen is much less in austenite than in ferrite and it is also more soluble in austenite). The cleanness of the steel can also be a factor, albeit a minor one, in that the inclusions and microvoids which result from impurity elements such as sulphur can act as 'traps' for hydrogen and thus reduce the effective hydrogen content [18]. Such micro inclusions are also helpful in another way in that they promote the nucleation of more desirable austenite transformation products and thus lessen the chance of martensite formation.

Control of stress: Since hydrogen induced cracking normally initiates at regions of stress concentration, smooth weld contours and good fit-up will help to prevent the occurrence of this defect (fit-up is one of the variables which is catered for in Appendix E of BS 5135).

Control of microstructure: Either the composition or the cooling rate through the austenite transformation temperature range must be controlled to limit the formation of martensite in the heat affected zone (or weld metal). These two factors are interlinked in that as the hardenability of the steel increases the maximum cooling rate to avoid the formation of a susceptible microstructure decreases. Therefore hardenable steels such as low alloy and creep resistant pressure vessel steels, particularly when in relatively thick sections, will need a significant degree of preheat (and sometimes even some post heat) to reduce the heat affected zone cooling rate to an acceptable level. The hardenability of conventional ferritic steels is determined by the following formula:

$$\text{Carbon equivalent (CE)} = \text{C} + \frac{\text{Mn}}{6} + \frac{\text{Ni} + \text{Cu}}{15} + \frac{\text{Cr} + \text{Mo} + \text{V}}{5}$$

The very great significance of carbon content has led to the development of constructional carbon and carbon–manganese steels of ever lower carbon content (e.g. the BS 4360 steels) to lessen the requirement for preheat and permit a wider range of welding procedures to be safely used. In such steels the strengthening effect of carbon is replaced by microalloying additions to give a finer grain size (e.g. BS 4360 normalised steels or control rolled steels) or by thermomechanical treatments to modify the austenite transformation (e.g. pearlite reduced or acicular ferrite line pipe steels).

In practice, the prevention of hydrogen induced cold cracking is normally obtained by a combination of methods—the use of low hydrogen consumables and preheat and heat input control together with, where possible, the use of low hardenability steels.

Chevron Cracking

A particular form of weld metal hydrogen induced cracking has recently become a significant problem. This type of cracking with its characteristic appearance (Fig. 3) is found in high strength ferritic steel weld metals and, although a hydrogen induced phenomenon, often originates at small hot tears [19]. The greater the weld metal strength and hardenability and the greater the tendency to hot cracking, the greater the potential problem. The best preventive measure is the reduction of the hydrogen level in the weld deposit (see Table 2).

Lamellar Tearing

This is not strictly a weld defect but a defect in plate which can be exposed by welding. The bonding between inclusions and the matrix in the base metal is weak and some of the inclusions will be brittle. Also these inclusions will tend to be elongated in the rolling direction of

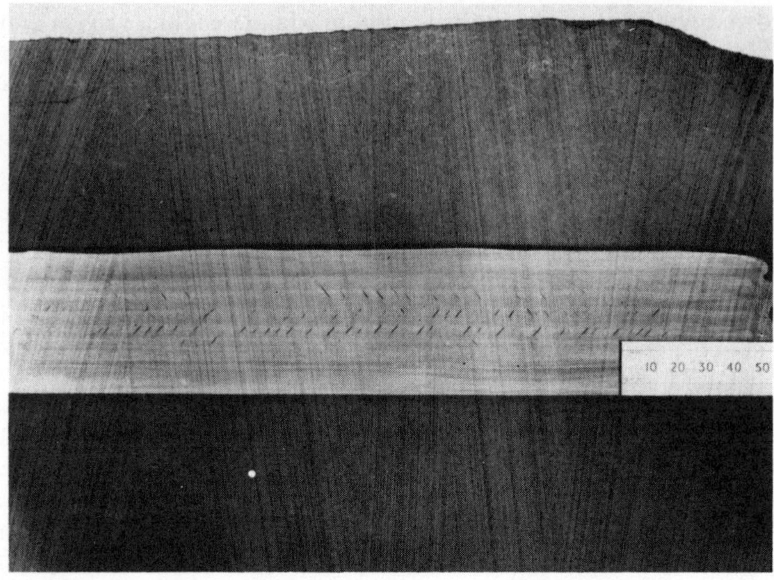

FIG. 3. Longitudinal section of a weld metal in a C–Mn steel showing chevron cracks.

plate. Therefore excessive strain in the through thickness direction can cause decohesion and fracture of inclusions leading, sometimes, to extensive tearing (see Fig. 4). A weld is often the cause of such excessive strain and although modifications in welding procedure and joint design to reduce such strains are possible preventive measures the most effective are those which improve the properties of the base

TABLE 2

EFFECT OF PRE-TREATMENTS ON HYDROGEN LEVEL AND CHEVRON CRACKING TENDENCY IN SUBMERGED ARC AND WELD METAL [19]

Flux treatment	*Weld metal diffusible hydrogen (ml/100 g)*	*Number of cracks per 100 mm of weld*
Baked at 450°C	2·5	0
As-received	3·7	2
Exposed to welding shop atmosphere for 10 days	6·2	10

FIG. 4. Typical lamellar tear adjacent to a large fillet weld [31].

metal. Therefore low inclusion levels (e.g. low sulphur steels) or modification of inclusions in terms of shape and ductility (e.g. rare earth treatments of steels) which improve the through thickness ductility (a minimum of 20 % reduction of area is recommended for ferritic steels [20]) are the best methods of preventing this problem arising.

Workmanship Defects

Solid Inclusions

The most important (and most frequently occurring) type of solid inclusion is slag inclusion. Such inclusions arise because it is difficult to ensure that all pockets of slag are removed from the relatively uneven surface of a weld, particularly when access is difficult. Fluxes do vary somewhat in the 'detachability' of the slags they produce, for example the older highly basic submerged arc fluxes are being superseded partly because of their poor slag detachability.

Oxide inclusions are occasionally found, particularly in welds in aluminium where they result from inadequate precleaning of the joint surfaces.

Tungsten inclusions are a defect associated with TIG welding where the use of an excessive current for a given electrode size or the chance touching down of the electrode into the weld pool cause the melting off of some of the electrode.

Porosity

Porosity occurs when the solid weld metal is supersaturated with a particular gas (hydrogen, nitrogen or carbon monoxide) which then forms pores as a result of the nucleation of gas bubbles on discontinuities in the metal (grain boundaries, micro inclusions, etc.). Since there will always be a sufficient number of nucleating sites the incidence of porosity in a given weld metal is a function of the degree of supersaturation of the relevant gas.

Gases enter the weld pool through air entrainment in the arc atmosphere (hydrogen and nitrogen), grease and moisture on joint faces or welding consumables (hydrogen) or chemical reactions in the weld pool or arc (carbon monoxide in steel).

Compositional and welding process and procedure factors primarily determine the range of porosity which can be expected. Examples of this are the effectiveness of the deoxidation reaction in carbon dioxide welding on porosity due to carbon monoxide, the nitrogen porosity due to air entrainment in welds made with gasless cored wires and the tendency to hydrogen porosity in MIG welding of aluminium because of the large surface to volume ratio of the electrode wire and the relatively low solubility of hydrogen in solid aluminium. However, the actual occurrence of porosity is a function of such factors as instability of the arc column due to incorrect electrode manipulation, inefficient cleaning of edge preparations, inability to control the arc column at stops and starts; all of which locally increase the gas content of the weld metal.

Lack of Fusion and Lack of Penetration

These defects are generally a function of electrode manipulation, joint design and arc current or inadequate preparation of the joint surfaces. The result is that the welding arc is not sufficiently 'penetrating' to 'wet' the edge preparation or the previously laid weld run or does not completely fill the joint gap.

As for porosity defects, there are certain process and procedure factors which make the defect more or less likely even though an individual occurrence is a workmanship error. For example, high heat input processes such as submerged arc or electroslag welding are not very prone to this defect whereas solid wire MIG processes are. Cored wire processes are less prone than solid wire processes because the arc reactions are frequently exothermic and provide a more efficient heat source.

Shape Defects

This range of defects (poor profile, undercut, misalignment, excessive spatter, for example) is almost exclusively a consequence of poor electrode manipulation or bad fit-up and the cause is frequently self evident. Sometimes the cause is an incorrect procedure. For example, undercut can be caused by too high a welding current or too low a speed and excessive penetration can result from too high a heat input. Excessive spatter may not be a result of poor electrode manipulation but can be, in the case of gas shielded welding, due to too low a current in the case of spray transfer or insufficient inductance in the case of dip transfer. In certain circumstances however incorrect procedures (wrong current or arc voltage or incorrect 'tuning' of the short circuiting carbon dioxide process) can be the cause.

THE SIGNIFICANCE OF DEFECTS

The fabricator's aim is to produce a welded structure without defects but, as discussed above, this is in practice difficult if not impossible if a structure is to be fabricated at an economic cost. Therefore all the major codes and standards which govern the quality of welded structures permit some latitude in this respect. In other words, they all define a tolerance level for weld defects. These tolerance levels have been defined on the basis of experience and engineering judgement and are to some extent arbitrary. They implicitly define for each type of construction (e.g. pressure vessel, pipeline, storage tank, etc.) the minimum quality standard which a competent fabricator should be consistently able to meet. From a quality assurance and quality control point of view this is a very appropriate way of setting the acceptable defect level but from a purely technical point of view which considers a given defect in terms of its effect on the integrity of the structure this is an inexact and possibly over conservative approach. The accumulation over the past 15 years or so of a great amount of data on the effects of weld defects on weld performance has shown that in this strict technical sense the acceptance standards in many of the major codes and standards are inappropriate and over conservative. This has led to a continuing debate on the viability of such acceptance standards and the development of formal methods, based essentially on fracture mechanics analyses, for assessing the significance of particular defects on a 'fitness-for-purpose' basis. (See pp. 130–1.)

The Effect of Defects on Weld Performance

From the point of view of their effect on the mechanical properties of welds we can classify defects into three main categories—volumetric (e.g. inclusions, porosity), planar (e.g. cracks, lack of fusion) and shape (e.g. undercut, misalignment) and we can consider their importance in quantitative terms on three types of loading or possible failure modes—static tensile, fatigue and fast fracture. Under this defect type classification we find that the majority of the planar defects will be technological defects whereas the majority of the volumetric and many of the shape defects will be workmanship defects. This has some bearing on the quality assurance and quality control implications.

Static Tensile Loading

Work on ferritic steels [4], aluminium alloys [21, 22] and copper alloys [23] indicates that we can assume, at least to a first approximation, a linear relationship between defect size and reduction in tensile strength. Volumetric defects by their nature cannot create a significant reduction in weld cross section (a weld with >5 % by volume porosity, for example, is almost too bad to be achievable) so that, in practice, we can ignore such defects on a 'fitness-for-purpose' basis in terms of static loading. Planar defects are more important because they can, in principle, cause a significant reduction in cross section (>50 % would not be impossible) and also because it is very difficult to accurately 'size' such defects in the through thickness direction with currently available NDT methods. Shape defects can also be significant (undercut of 10–15 % is not unknown) but the size of these defects can at least be accurately measured.

Fatigue Loading

Fatigue cracks originate from 'notches' which produce a stress concentration under an applied stress. A welded joint in itself generates a stress concentration, the magnitude of which varies considerably with the joint design. This being so the fatigue strength of a welded joint is highly dependent on the joint design (for example, a sound butt weld in carbon steel stressed transversely will have a fatigue strength for a given endurance some five times greater than a sound non-load bearing fillet welded attachment in the same material) and construction codes such as BS 5400 Steel, Concrete and Composite Bridges and CP 118, 1969 The Structural Use of Aluminium sensibly take this into account. Also, since the majority of the fatigue life of a welded joint consists of crack propagation rather than crack initiation and crack propagation rates are insensitive to microstructural variations and therefore to compositional changes the major variable which determines the fatigue performance of a welded joint is the stress concentration resulting from the joint design. This being so, the significance of any weld defect is determined by the degree to which it increases the already existing stress concentration. A given defect will therefore not necessarily have the same importance in welds of different types and, furthermore, the position of the defect within the weld cross section can be a critical factor.

Since weld shape is so important in determining the fatigue performance shape defects and surface breaking defects are generally the most serious. For example, in butt welds the fatigue strength can be related to the reinforcement angle. Volumetric defects can usually be ignored as they will have no significant effect on the stress concentration except in some very special cases [24] where the inherent stress concentration is very low and there are surface breaking volumetric defects.

The presence of planar defects can obviously result in an increase in stress concentration, the effect being greater if such a defect is at a weld toe and being relatively more serious the greater the inherent fatigue strength of the joint design. For these reasons the setting of weld quality standards where fatigue is the operative failure mode must take into account the joint design, e.g. butt welds will need to be fabricated to a higher standard than fillet welds to avoid a degradation in fatigue strength. This is taken into account in the more enlightened structural codes (e.g. BS 5400 and CP 118). (See Chapter 2.)

Fracture

Fracture mechanics analyses demonstrate that fracture can initiate from a defect when under load if the stress intensification at the defect is sufficiently high for it to be energetically favourable for a crack to initiate. This implies a relationship between the defect size and position and the magnitude of the applied stress (the factors which determine the degree of stress intensification) and the toughness of the weldment in the vicinity of the defect (the factor which determines the critical level of stress intensification for a crack to initiate). This inter-relationship between a number of factors (not all of which can be accurately measured in many cases) means that the assessment of the significance of a defect in terms of fracture must be undertaken by a fracture mechanics analysis and standardised analytical methods are available for this (see next section). However, some general statements can be made. Volumetric defects are, again, of little significance in that because of their shape they cannot generate a sufficient stress intensification to be harmful in the materials normally used in welded construction. Shape defects, likewise, are unlikely to be serious defects. Planar defects are the most significant and the stress intensification resulting from such a defect will be a function of the applied stress and the defect size. Because of the difficulty in accurately measuring the size of the planar defects it is almost always necessary to consider planar defects as unacceptable if fracture is a possible failure mode unless the material toughness is extremely high in relation to the defect size.

Formal Methods for Assessing the Significance of Defects

The philosophy behind all the formal methods which have so far been introduced was propounded in two papers published by Harrison, Burdekin and Young in the late 1960s [25, 26]. Modifications to the details have been made subsequently (for example, Harrison [27, 28, 29]) and a critical review of the different methods has been produced by Burdekin [30].

In practice, the formal methods are concerned primarily with the possibility of a defect initiating fatigue failure or fast fracture and a defect is considered 'acceptable' if it can be shown by a fracture mechanics analysis that the possibility of failure occurring from the defect is remote. Since a fracture analysis of this type relies on making assumptions or best estimates of many of the parameters (stress distribution, flaw size, fracture toughness, fatigue crack growth rate, residual stress, for example) which are difficult to measure precisely there are bound to be differences between the different methods which relate primarily to the degree of conservativeness (or factor of safety) which is included in the calculations. Clearly, therefore, the more precise the design and material property data the more useful and reliable such fitness-for-purpose analyses can be. For example, the approach used in the ASME Boiler and Pressure Vessel code Section XI Appendix A for evaluating flaws found during in-service inspection of a nuclear component is very exact. The properties of the steels used are known with some precision as are the environmental effects (e.g. irradiation) on them. Very detailed design analyses will be available to cover all expected operating modes (including 'fault' and 'emergency shutdown' conditions) and, furthermore, the inspection techniques and hence the reliability of the flaw size estimate will be good. Even in this case a very large factor (10 on the flaw size necessary to cause failure under normal operating conditions) is used to estimate the tolerable flaw size and all flaws are considered in the same way there being no distinction between volumetric and planar defects.

The British Standard document PD 6493:1980 Guidance on Some Methods for the Derivation of Acceptance Levels for Defects in Fusion Welded Joints gives a method of analysis for all types of structure and is necessarily, therefore, less exact. It only requires an analysis of planar defects as volumetric defects are assumed to be insignificant unless the toughness is low ($<1300\,\mathrm{N\,mm^{-3/2}}$). A limit is set for volumetric and other minor defects which is a somewhat arbitrary limit based largely, one suspects, on what is thought to be an 'acceptable' quality level. Since the document can only give general rules because the amount of, and precision of, the relevant engineering data will vary considerably according to circumstances the document is perhaps not as widely useable as would be expected. Its major use, and that of other, similar, standards, in fact, may well be in material and welding process selection. If it can be assumed that for a given structure defects above a given size will either not occur or will be detected with a very high reliability then by making conservative assumptions about maximum stress levels and defect positions it is possible to calculate the minimum fracture toughness required of the materials and weldments so that fatigue or fracture are unlikely to occur. This fracture toughness level can then be imposed as a material specification and will be of major importance in the quality assurance of the structure.

CONCLUDING REMARKS

Weld defects are, obviously, undesirable by definition but the complexity of the technology of welding means that it is not usually a worthwhile aim to expect total freedom from all defects.

The major, technological defects (cracks and other planar defects) can and should be prevented to a large extent by proper quality control over materials, joint designs and welding procedures as the causes of such defects are well known.

The minor, workmanship defects are not totally preventable and a degree of tolerance for them is appropriate. Such tolerance levels are based largely on experience of what is achievable by competent fabricators and no significant change in this approach seems likely or, from a quality assurance point of view, desirable.

The analysis of the significance of defects supports what would intuitively be expected, the crack and crack-like defects are the most serious and the volumetric (porosity and solid inclusions) defects are of little importance in terms of structural integrity. However formal methods of assessing crack-like defects in terms of their significance in relation to possible structural failure, whilst correct, are difficult to apply in many cases because of the lack of sufficiently detailed information on defect size, stress level and fracture toughness. These formal analytical methods, though, are of considerable importance from a quality assurance sense in helping to specify and characterise material property requirements.

REFERENCES

1. Classification of defects in metallic fusion welds with explanations. *Welding in the World,* **7** (1969) 200–10.
2. Parameter characterising defects in metallic fusion welds. *Welding in the World,* **9** (1971) 92–111.
3. Salter, E. R. and Gethin, J. W., 'Pressure Vessel Standards—The impact of change'. The Welding Institute, London 1972.
4. Kihara, H., Tada, Y., Watanate, M. and Ishii, Y., Non-destructive testing of welds and their strength, Society of Naval Architects of Japan, 60th Anniversary Series **7** (1960).
5. Rogerson, J. H., Quality assurance in welding construction, International Institute of Welding, Public Session, Estoril, 1980.
6. Rodrigues, P. E. L. B., Figueiredo, R., Moura, I. and Guerreiro, M., Quality assurance in welding construction, International Institute of Welding, Public Session, Estoril, 1980.
7. Rogerson, J. H., The development of efficient and rational sampling schemes for the NDT of welded fabrications, *Weld. Res. Int.,* **7** (1977) 412–33.
8. An Assessment of the Integrity of PWR Pressure Vessels. Report by a study group. UKAEA HMSO, 1976.
9. Rodrigues, P. E. L. B., Wong, K. H. and Rogerson, J. H., Paper 3693 Offshore Technology Conference, Houston, 1980.

10. Becher, P. E. and Hansen, B., Statistical Analysis of Defects in Welds, Danish Welding Institute, 1974.
11. Borland, J. C., Generalised theory of supersolidus cracking, *Br. Weld. J.,* **7** (1960) 508–12.
12. Borland, J. C., Suggested explanations of hot cracking in mild and low alloy steels, *Br. Weld. J.,* **8** (1961) 526–40.
13. Simpson, M., Solidification cracking during the submerged arc welding of carbon–manganese steels—a review, *Weld. Res. Int.,* **7** (1977) 177–92.
14. Hemsworth, B., Boniszewski, T. and Eaton, N. F., Classification and definition of high temperature welding cracks in alloys, *Metal Construction,* **1** (1969) 5–16.
15. Young, J. G., Significance of filler metal composition in aluminium alloy welding, *Br. Weld. J.,* **8** (1961) 568–74.
16. Nichols, R. W., Reheat cracking in welded structures, *Welding in the World,* **7** (1969) 244–60.
17. Dhooge, A., Dolby, R. E., Sebill, J., Steinmetz, R. and Vinckier, A. G., A review of work related to reheat cracking in nuclear reactor pressure vessel steels, *Int. J. Press. Vess. Piping,* **6** (1973) 329–409.
18. Hart, P. H. M., Paper 20 Trends in Steels and Consumables for Welding, The Welding Institute, London, 1978.
19. Mota, J. F. *PhD Thesis,* Cranfield Institute of Technology, 1979.
20. Wilson, W. E., Minimising lamellar tearing by improving 2-direction ductility, *Weld. J.,* **53** (1974) 691–5.
21. Dinsdale, W. O. and Young, J. G., Significance of Defects in Aluminium Fusion Welds, 2nd Commonwealth Welding Conference, London, 1965.
22. Lancaster, M. V. and Rogerson, J. H., Welding in Non Ferritic Materials, The Institute of Welding, London, 1967.
23. Hudson, M. E., Mota, J. F. and Rogerson, J. H. Cranfield Institute of Technology, unpublished work.
24. Dawes, M. G., Fatigue strength of ferritic steel shafts reclaimed by welding and metal spraying, *Br. Weld. J.,* **10** (1963) 418–38.
25. Burdekin, F. M., Harrison, J. D. and Young, J. G., 'The Significance of Defects in Welds', The Institute of Welding, London, 1967.
26. Harrison, J. D., Burdekin, F. M. and Young, J. G., A proposed acceptance standard for welded defects based on suitability for service, 2nd Conference on the Significance of Defects in Welds, The Welding Institute, London, 1968.
27. Harrison, J. D., Basis for a proposed acceptance standard for weld defects—Part 1: Porosity, *Metal Construction,* **4** (1972) 99–107.
28. Harrison, J. D., Basis for a proposed acceptance standard for weld defects—Part 2: Slag inclusions, *Metal Construction,* **4** (1972) 262–8.
29. Harrison, J. D., A re-analysis of fatigue data for butt welded specimens containing slag inclusions, *Weld. Res. Int.,* **8** (1978) 81–101.
30. Burdekin, F. M., Colloquium on the Practical Application of Fracture Mechanics, The International Institute of Welding, Bratislava, 1979.
31. Ferreira dos Santos, A. C., *PhD Thesis,* Cranfield Institute of Technology, 1978.

6

The Inspection of Welds and Welded Construction

N. T. Burgess

Gilbert Commonwealth Engineers and Consultants (UK) Ltd, Twickenham, UK

INTRODUCTION

Inspection is a part of quality assurance and this chapter identifies some aspects of the work required. Fortunately, inspection activity is the subject of many standards around the world to which reference should be made for detailed guidance in specific industries or instances [1–5].

GENERAL

As long as human beings are involved in welding operations, as welders or as machine operators there will be variability in performance and some inspection of their work will be required.

It was earlier hoped that the move to automatic and semi-automatic welding would improve on the quality of manually made welds and that the need for post weld inspection would decrease. This has not proved to be the case and, indeed, some feel that new developments in welding technology, material and in design have themselves contributed to welding problems by introducing complications that may not have been assessed for the effect on quality assurance. In Chapter 3, A. Gifford gives such examples.

Whilst it is true that quality control principles can be more readily applied to machine welding than to manual welding, we remain heavily reliant on the 'inspector' to determine whether welds meet the specified requirements or not. Inspection activity, however, must be seen as part of the total quality control effort and not as an end in itself. It must be programmed into the total scheme of checks since inspection alone as the main QA method would be too late to affect the quality of the product.

Inspection and NDT are often regarded as synonymous but this is not the case. Whilst NDT is a major tool in controlling welding quality, its contribution must be kept in perspective and reliance upon it treated cautiously (Chapter 7).

We must also be clear as to which party is best fitted to carry out the inspection of welds. Too often in welded construction inspection is left to the customer's inspector or to a third party or official inspectorates, although under the principles outlined in Chapter 1, it is primarily the manufacturer's responsibility to ensure that his welds are to specification. The initial inspection activity, therefore, must take place at his behest, in his time and on his premises. If his inspections are carried out effectively then there may be little or no need for further inspection by customers or third parties.

For such inspection activity to be effective, several factors must be considered: the skills of the inspector, the general inspection training he has received, the equipment he has (including his eyes) and the information and briefing he has received about specific weldments. Those involved with inspection and acceptance of welds require skills that span the total range of capabilities.

Although, as has been shown, the prime responsibilities rest with the shop (or site) inspector of the organisation making the weldment, there is a reluctance among manufacturers to recognise the need to adequately support this responsibility with the right people (or to have them approved). This is noted by customers and the authorities and visiting inspection engineers (insurance companies, classification societies, etc.) who generally carry out 'witness inspection' may bring further knowledge into decisions concerning acceptance or rejection. It is felt that welding is so important that nothing but a specialised knowledge of the likely problem areas will suffice. However, manufacturers must be wary of the 'expert' overseer who, for example, may wish to interpret radiographs. This activity may look easy, but everyone including the 'expert' should be willing to demonstrate his expertise by being qualified in the skill.

Responsibility of Welding Inspectors

Inspection of welds should be carried out by qualified and specially trained personnel whether working for the manufacturer or the customer. In the oil, gas and structural industries, welding inspection is often recognised as a particular discipline, whereas in other industries, inspection of welds is not specifically identified and the work is done as part of general inspection by mechanical inspectors and engineers. Welding is not picked out from machining, assembly etc. but as long as the individual recognises the importance of welding variables, and is trained and qualified, there should be no problem.

Many countries operate schemes for the certification of welding inspectors, examples being:

(a) UK—Certification Scheme for Weldment Inspection Personnel (CSWIP)
(b) USA—AWS Welding Inspector Qualification and Certification Scheme AWS QC1-76
(c) Canada—Qualification Code for Welding Inspection Organisation
(d) Australia—SAA Welding Certification Code, AS1796-1975 and SAA Structural Steel Welding Supervisors Certification Code, AS2214-1978.

The UK scheme [6], which is similar to the others, places specific responsibilities on the welding inspector as follows:

(1) *Codes and standards:* Interpretation of the requirements of codes and standards.
(2) *Welding procedures:* Establishing that a procedure is available, has been approved by the appropriate authority and is being employed in production.
(3) *Witnessing of welder and procedure approval tests:* Witnessing the preparation of test plates and destructive tests and verifying compliance with appropriate standards and specifications.
(4) *Welder approvals:* Verification that adequate and valid welder approvals are available, and that only approved welders are used in production.
(5) *Parent material identity:* Verification against documentation and markings of correctness of parent material.
(6) *Welding consumables identity:* Verification of correctness of welding consumables (electrodes, filler wires, consumable inserts, gases, fluxes etc.)
(7) *Pre-weld inspection:* Verification that dimensions, fit-up and weld preparations are in accordance with specifications.

(8) *Preheating:* Verification that preheat (where required) is in accordance with specified procedure.

(9) *In-process welding surveillance:* Surveillance during welding to verify compliance with specified procedure including any pre-heat, interpass temperature control and post heat requirements.

(10) *Post-weld heat treatment:* Verification that post weld heat treatment has been conducted in accordance with specification requirements.

(11) *Post-weld visual inspection:* Visual inspection and dimensional check of completed weldment against specification requirements and drawings.

(12) *NDT reports:* The study and cognisance of NDT results on any welding work for which the welding inspector is responsible. If the duties of the welding inspector include the interpretation of weld radiographs it is suggested that he seeks certification in accordance with the related approval schemes.

(13) *Reports:* Preparation of inspection reports.

A welding inspector with responsibilities and capability to meet the above will bring great benefit to the QA programme. To be accepted under the UK CSWIP scheme for welding inspectors, candidates must have had at least three years experience of the duties required, under qualified supervision. Successful completion of courses on welding inspection may qualify for a reduction in the period.

For certification of a welding inspector, approval consists of written, oral and practical examinations. For the guidance of candidates and their employers, a specimen written examination paper and syllabus is provided, which outlines the subjects covered in the examination. (The Welding Institute of the UK can provide further details.) The following summarises the main points.

Written Examination

The written examination is designed to test the candidate's knowledge of welding processes, procedures and their control, welder approval, defects and their origin, heat treatments, welding consumables, weldability of materials (as appropriate), destructive tests, terminology for welds, welded joints and weld defects, standards and codes of practice, capabilities of NDT methods, visual examination and dimensional checking, reporting. That part of the examination concerned with the interpretation of codes and standards will be of the 'open book' type and candidates must take a copy of the standard with them.

Oral Examination

The oral examination is used to supplement the written examination and covers the same subject matter. It normally consists of a discussion with the examiner during the practical tests.

Practical Examination

Candidates are required to inspect and report on the following:

(i) at least two completed welds for compliance with stated requirements;
(ii) a set of destructive tests (including macros) for a welder or procedure approval examination intended to comply with a stated specification.

Similar requirements are being laid down in more and more countries and increasingly customers are specifying that people employed on their work should be so certified.

Stages of Inspection

In general, inspection can be:

(a) *Final Inspection* of the completed component. In the case of welded joints this may include the witnessing of performance tests, e.g. pressure test if a vessel is involved or a load test for structure or moving components.
(b) *Stage Inspection,* the normal activity for welded structures, in which the inspector's activity will commence as early as the design stage with checking of weld procedures against the specification (see the section headed 'Responsibility of Welding Inspectors') and continue to final inspection. (Stages are usually prescribed.)
(c) *Patrol Inspection,* most common in general engineering workshops, involving a combination of surveillance of production operations against instructions, physical inspections and examinations.

Which Party Should Do the Inspection?

Welded constructions have always been subject to a great deal of inspection by customers and third parties, in addition to the inspections that are carried out by the manufacturer or contractor himself. It should be recognised, however, that the most effective inspection is generally that carried out by the manufacturer, since he is closest in all respects (physically and commercially). Unfortunately, even in well regulated systems, with certified staff, mistakes do occur, often due to human error and customer activity is still common. This can be moderated to surveillance and monitoring of the manufacturer's QC action when this is more effective. A quality assurance policy introduces this possibility of quality surveillance, which can be considered as a comprehensive version of patrol inspection (c) (above) applicable to the case of welded structures by involving (a) and (b) also.

With properly set up quality control systems (see Chapter 1) surveillance or monitoring on the part of customers, may be all that is necessary to assure quality. 'Witness' inspection may however be worthwhile where it forms part of a planned system for monitoring a manufacturer's QC system rather than for 'acceptance or rejection'.

Aids to Inspection

Acceptance of welded joints (other than by NDT) relies generally on visual inspection, for which one individual and the human eye are the main tools. 'Nothing can totally replace ever-open eyes, a sharp pair of ears and a quick pair of hands, all trained to act in response to the well programmed computer we all have between the ears'. Good eyesight, with the aid of properly focused glasses if necessary, is essential. Other useful equipment would be a magnifying glass; a torch for illuminating parts of the weldment not properly illuminated and a wire brush for removal of rust, slag, etc.

Illumination

Natural daylight is clearly beneficial to good inspection, but this may not always be possible. The effectiveness of illumination largely depends upon contrasts. The area to be examined should be adequately and evenly illuminated without shadow or glare. A general guide is that illumination should not be less than 500 lx. If good daylight is not available, then appropriate artificial lighting must be provided. Hand lamps can be useful for localised inspection. Fluorescent tubular lighting can largely eliminate shadow but may introduce some undesirable flicker.

Surface Cleanness

The surface of welds to be examined must be clearly exposed. This may require the removal of slag, rust, oil or other extraneous dirt, etc. (Inspectors should beware of being asked to inspect/accept weldments that have been painted). Methods normally used for cleaning the material under examination may be used except any that could mask the feature for which examination is to be made. Some mechanical methods may tend to close discontinuities.

Dressing of Welds

This is a contentious subject, earlier the cause of trouble, nowadays regulated but nevertheless deserving of caution. Welds are dressed to improve shape and profile and therefore fatigue characteristics (see Chapter 2). Dressing to improve appearance or to clean up may be deleterious in masking defects, or, in the risk of over-flushing, may also be costly. Scratch brushing may be adequate in many cases. Where surface crack detection is specified preliminary dressing may be required. In all cases the QC specification should be clear on these points.

Other Aids

The sense of feel can be a useful adjunct to visual examination, e.g. the use of a pin to confirm or explore a crack or other surface defect. A straightened paper clip can be even more sensitive in that it is softer and so can bend, e.g. when inserted to assess the depth of a pore.

A common aid to vision is an ordinary low power pocket lens of magnifying power in the range of ×2 to ×10. The range of optical devices for the effective study of positions inaccessible to the eye is steadily increasing. Some using mirrors and lenses (e.g. intrascopes, borescopes) have been established for many years; others such as light guides and miniature television cameras are more recent developments. All can be very effective when appropriately used.

Dimensional and Shape Checks

The Welding Institute have produced an effective gauge for establishing the weld profile and other characteristics and this is particularly useful with fillet welds (Fig. 1). The gauge will determine fillet weld leg length, misalignment, throat size, depth of undercut or pitting and excess weld metal.

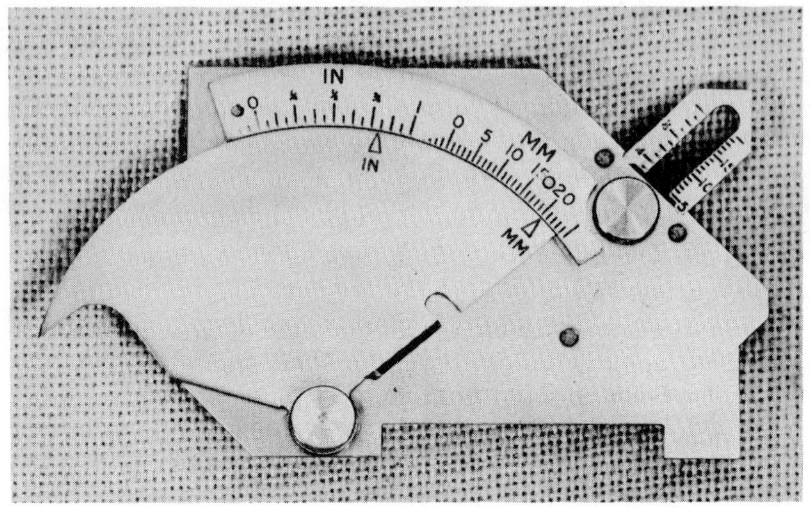

FIG. 1. Multi-purpose welding gauge.

Human Factors in Inspection

We cannot eliminate human error or change human nature and even the decision makers need to make decisions about *which* decisions to make!

In mass production industries, automatic inspection is replacing much human inspection since a machine is generally more efficient in a 100 % checking situation of a continuous flow line. But when production operatives and production conditions are variable, as with welding, it is difficult to find a machine capable of examining for the *several* different characteristics or for such indefinite attributes as 'finish'.

It is necessary to make the best use of conditions for the inspector. The factors that affect his judgement are: working environment, temperature, noise and visual environment. Adverse conditions can reduce efficiency, produce physical discomfort or cause damage to eyesight (e.g. arc flash). As noted earlier, an important factor is the amount of lighting available for inspection.

Perhaps the most important factor however is the 'social' one. Fortunately, in welding constructions, the inspector is held in fairly high regard compared with his colleagues in the machine shop. This is probably the result of third party influence as noted earlier and also the

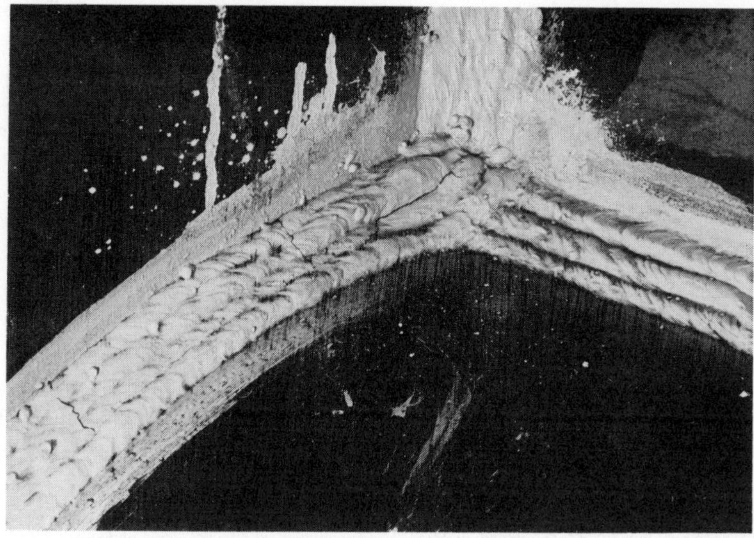

FIG. 2. Photograph of fillet weld finish (containing crack) showing how photographs can be used as reference standards.

number of technical aspects associated with welding. For shop or site inspections by the manufacturer/contractor, the relationships between inspector and production worker (welder) is critical. The welder may see the inspector as the person who aims to stop him producing the volume of work required. It may tempt the inspector to give concessions in order to remain friendly; equally an adverse relationship may cause friction to a degree that the inspector is unfairly rejecting items for trivial faults. Much welded work is judged subjectively rather than objectively since some of the standards are unavoidably difficult to specify on paper. NDT has introduced a quantitative element but evaluation of, for example, surface profile and finish remains a problem and one that has cost many welding fabricators dearly in the past. This is due to reliance on the subjective views of the visiting inspector who may have a very personal view on what is acceptable. A clear company QC standard in this area would help. Plastic and rubber replicas of acceptable/unacceptable weld finishes have been used in the past. Photographs are also valuable. (Fig. 2)

A more general but very important point is that if the inspector

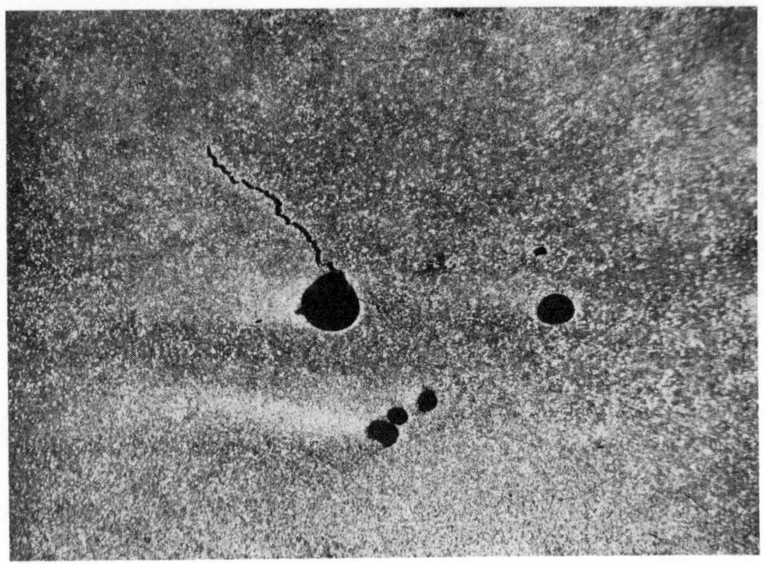

FIG. 3. Cracking associated with, and probably masked by porosity.

(inside or visiting) spends too much time on one characteristic, particularly one that is *easy* to inspect, he may overlook other more important characteristics or defects. An example arises with the radiographic examination of welds in relation to porosity. Because porosity is readily apparent on most radiographs, it is easy to identify and comment upon, and the inspector may take a lot of convincing that large pores can be admitted as acceptable. Detailed discussion will ensue concerning size, depth, location, etc. before ultimate rejection, repair or acceptance. However, many crack types (see Chapter 5) may not so easily be seen on the same radiographs! There is no debate, and they may be overlooked in the argument concerning porosity. Figure 3 is a photograph of blowholes with associated (and potentially more serious) cracks that were missed in the radiographic report. It is now generally accepted [6] that quite large amounts of porosity may have little affect on the strength of welds, whereas quite small cracks and fissures most certainly do.

Guidance to Welding Inspectors—Fitness for Purpose and Acceptance Criteria

Figure 4 illustrates how the acceptance standard for products may be capable of adjustment to suit the needs of the design requirement rather than that of an arbitrarily written code. Where material characteristics of the construction are known (particularly the fracture toughness) and where NDT results can be accurate, 'fitness-for-purpose' levels of defects B can be determined and used for accept or reject decisions. This level is the one below which welds cannot be accepted (at least without repairs, etc.). Above this level (i.e. less defects) welds may be acceptable but if they contain defects greater in quantity or seriousness than level A welding operations must be subject to improved quality controls so as to bring quality standards up to the specified or quality control level.

A common sense approach is necessary because research work has shown that, for example, the tolerance of mild steel weld metal to spherical defects such as uniformly distributed porosity, is so high that the acceptable level of porosity would probably prevent effective non-destructive testing of the seam from the point of view of the detection of more harmful defects, such as cracks, lack of fusion and lack of penetration! It follows that once the scientifically established (fitness-for-purpose) criteria are laid down, such criteria can only be used for those fabrications where 100 % NDT is carried out [6]. However, more and more organisations are using this approach.

For the monitoring of workmanship quality levels, i.e. using NDT during the fabrication process rather than for acceptance, it will still be

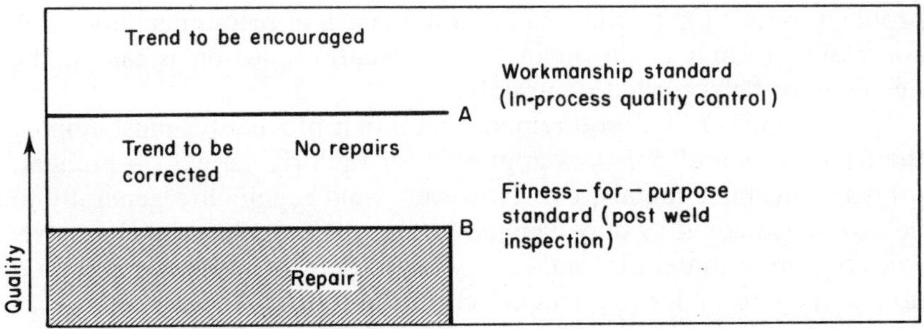

Fig. 4. Fitness for purpose/workmanship standards.

necessary to use arbitrary levels of weld quality such as those appearing in many codes. As information on the significance of defects grows, these arbitrary levels of weld quality can be modified and set at suitable values. Of course, downward trends in quality must be corrected by attention to the fabrication process before the risk of poor quality becomes significant. Unfortunately, at the present time too many fabricators feel that it is cheaper to carry out repairs, however unnecessary, than to argue with the visiting inspector since argument itself can involve delays to production and delivery dates which maybe even more expensive than the repair itself:

REPAIRS

Apart from the 'black and white' accept or reject possibility there are various other options available to the manufacturer. These can be classified as follows:

1. Repair/rectify—remove the defects and reweld. Concession necessary.
2. Replace—the component. Concession may be necessary.
3. Rework—e.g. dress the weld to remove minor defects—then accept 'as is'. Concession probably necessary.

In most cases a quality assurance system will require that a concession procedure operates. This may involve the manufacturer or the inspector raising a report (concession request) to the designer, then customer, or some other party, which identifies the defect, states the problem requiring a concession and makes a recommendation. A concession format is shown in Fig. 5. Whatever action is taken, the details *must* be recorded accurately.

In most codes it is a requirement that repair procedures must employ the techniques and practices approved for the original work—preheat, preparation, filler metals, etc. However, weld repairs are generally to be avoided, since they may damage further the metallurgical characteristics of parent materials, and so any latent defects in the original weld may propagate under the heat effects of the repair. An investigation of boiler tubes failure by the UK power industry in the 1960s revealed that a large proportion of leaks occurred at welds that had been repaired. Several standards, e.g. BS 5289 : 1976, give additional advice

Sheet ______ of ______

CONCESSION REQUEST

CLIENT ______________________ DATE OF VISIT ________ REPORT NO. ______

CONTRACT NO. ______________________ ORDER NO. ________

VENDOR ________________ VENDOR REF. ________ CONTACT ________

ADDRESS ______________________

____

____

PROMISED DATE ______________ LATEST PROM ________

DESCRIPTION: DRAWING NO. ________

Details of Deviation

Proposed Disposition/Rectification Procedure

Effect on Delivery:

Effect on Cost/Manhours;

Originator Date Vendor Date

	Quality Assurance	Date		Project Manager	Date
1. ACCEPTABLE UNACCEPTABLE			2. ACCEPTABLE UNACCEPTABLE		
	Engineering	Date		Client	Date
3. ACCEPTABLE UNACCEPTABLE			4. ACCEPTABLE UNACCEPTABLE		

Form No:
Issue :

FIG. 5. Typical format for concessions.

on inspection of repairs. For example:

(a) Removal process—ensure that the specified means of removing the defect, e.g. chipping, grinding, machining, thermal cutting or thermal gouging is used correctly. When a thermal process is employed check that if preheating is specified it is correctly applied.
(b) Partially removed weld—check that the cut out portion is sufficiently deep and long to completely remove the defects. Ensure that at the ends and sides of the cut there is a gradual taper from the base of the cut to the surface of the weld metal, the width and profile of the cut being such that there is adequate access for re-welding.
(c) Completely removed weld—when a cut has been made through a faulty weld and there has been no serious loss of material, or when a section of material containing a faulty weld has been removed and a new section is to be inserted, check that each weld preparation is re-made in accordance with the welding procedures.

INSPECTION SAMPLING SCHEMES

Where large numbers of welds are involved in a project, or when a construction involves a large number of welds, a sample only is often proposed for inspection and acceptance purposes. If 100 % inspection (or NDT) is specified it must be made clear whether this means 100 % of each and every weld, or one inspection (e.g. one radiographic shot) of each weld. '100 % inspection' is a dangerous expression and largely meaningless without specifying which method of inspection should be used (i.e. visual or NDT) and to what standard it should be taken (i.e. what are the criteria for acceptance?). Beware also of simple sampling statements such as '10 % radiography'. Which 10 % should it be? 10 % of each working shift, or 10 % of all welds per day, or of each welder's work, or on each component? Should butt welds, which are relatively easy to examine by NDT, be covered to the exclusion of fillet welds or does the percentage apply to both butts and fillets? What about the attachment welds? It is for such reasons that each construction needs inspection procedures as part of the quality/inspection plan.

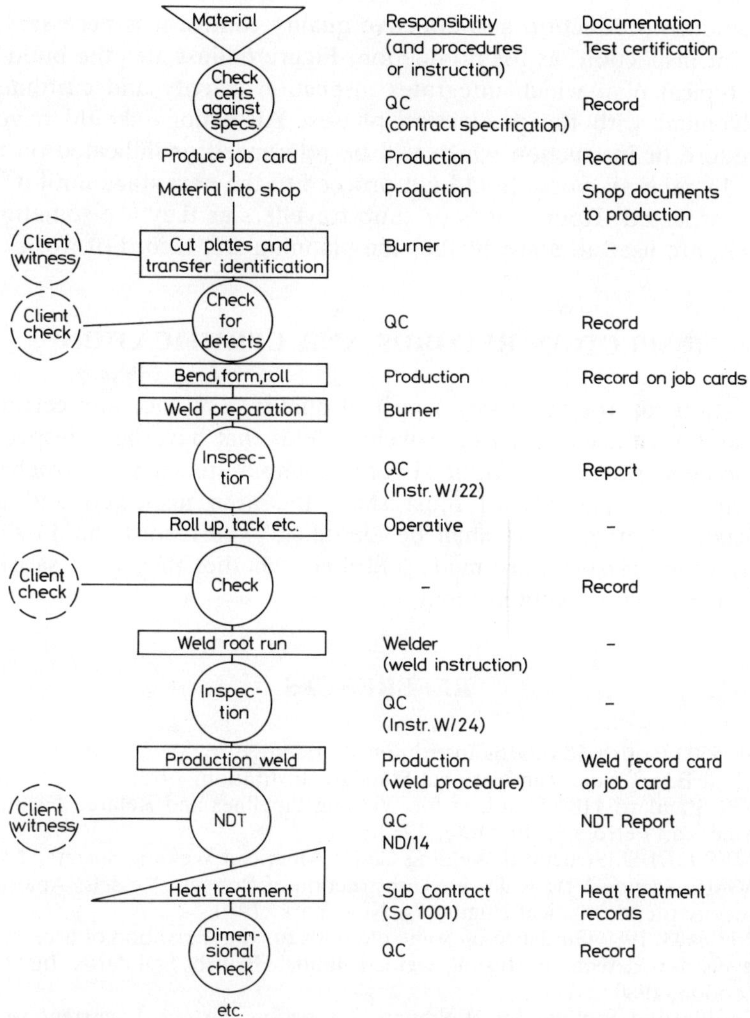

FIG. 6. A typical inspection plan. Symbols used: inverted triangle, input or supply; full line circle, QC check; rectangle, works operation; right angle triangle, sub-contract or special processes; dashed circle, client check or hold point.

THE INSPECTION PLAN

For smooth production and effective quality control it is necessary to plan for inspection, as for production. Figure 6 illustrates the build up of a typical plan which integrates inspection activity and customer's involvement with the production phases. All actions should have a procedure or instruction which will be referenced or indicated on the plan. In general, work should not proceed to the next stage until it has been inspected. Route cards or shop travellers as they are sometimes known, are used to assist further the planning and record of work.

INSPECTION RECORDS AND CERTIFICATION

Inspection records are a key aspect of quality assurance and certification and must accompany each weld. Welds that have been inspected should be marked and identified by, for example, stencil or electrochemical marking. The record must show the area inspected and any defects or characteristics shall be identified. The record should show whether any repairs were made and if so, whether they were satisfactory or not, who accepted them, etc.

REFERENCES

1. BS 5500 British Standards Institution, London, 1976.
2. ANSI B31.3 American National Standard Institution, 1973.
3. API Standard 1104 Standard for Welding Pipelines and Related Facilities, American Petroleum Institute, 1980.
4. AWS DI.1.80 Structural Welding Code, American Welding Society, 1981.
5. ASME Section VIII Rules for Construction of Pressure Vessels, American Society for Mechanical Engineers, New York, 1980.
6. P.D.6493 : 1980 Guidance on some methods for the derivation of acceptance levels for defects in fusion welded joints, British Standards Institute, London, 1980.
7. Certification Scheme for Weldment Inspection Personnel, current issue', The Welding Institute, Cambridge, UK.

7

Non-Destructive Testing of Welded Joints

T. J. Jessop
The Welding Institute, Cambridge, UK†

INTRODUCTION

It is important to consider NDT as an integral part of the whole fabrication process. It has always been necessary to urge designers to ensure that welds can be made to the required standard and at minimum cost; it is now being realised that design consideration should also be given to the NDT requirements and how they are to be met. It is essential that welding engineers and designers alike should be au fait with NDT methods to ensure the most effective quality control.

The requirement by the fabrication industry of maximum integrity at minimum cost, assisted by developments in welding technology and fitness-for-purpose considerations, has placed increasingly stringent demands on the capability of NDT methods. Indeed, the current problems with ultrasonic testing [1] would indicate that recent developments, particularly in fracture mechanics, which have enabled critical defect sizes to be quantified, are in advance of the present NDT capability. Some NDT advances have been made, and the fabrication industry should be aware of these and should consider the possibility of commercial exploitation. NDT methods are, however, greatly in need of further development and improvement, although much has been accomplished in recent years in the way of establishing links between companies experiencing similar problems and the research staff involved in solving them [2, 3]. This collaboration has provided a means of ensuring that research work is pertinent to industry's problems and has facilitated practical application of newly developed techniques.

† Present address: The Welding Institute, North American Office, PO Box 5268, Hilton Head Island, South Carolina 29938, USA.

This chapter is designed to provide information on all the above aspects. The NDT methods (and their limitations) commonly used for weld inspection are described together with a summary of recent developments and applications. For convenience, the methods have been divided into two categories: those capable of detecting surface flaws only, and those capable of detecting internal flaws. A comprehensive list of worldwide NDT standards appears in the Appendix.

SURFACE FLAWS

Welding defects such as undercut [4] can still be most conveniently detected by simple visual examination and this NDT method should never be ignored (see Chapter 5). Smaller defects such as surface cracks, however, require an increase in visual contrast between the defect and its background to be detected; this may be achieved by either the magnetic particle or the dye penetrant method. Both these methods are relatively inexpensive. Eddy current testing is also capable of detecting surface defects but may be difficult to apply. All these methods have the limitation that defect depth sizing is not possible.

Magnetic Particle Method [5]

This method is based on the principle that a discontinuity in a magnetic field causes flux leakage which strongly attracts magnetic particles because of the magnetic poles set up at the site of the defect. Surface defects and those lying just beneath the surface may be detected by this method. The magnetic field is set up in the test specimen either by passing current through it (for example by means of hand held prods), placing it in a current-carrying coil, or using a powerful permanent magnet or electromagnet. A looped conductor technique may have advantages for more complex joint geometries such as those found in offshore structures [5]. The magnetic particles are normally held in suspension in a suitable liquid and sprayed on to the test specimen while the magnetic field is applied. The direction of the magnetic field depends on the magnetising method and for greatest sensitivity the field must be applied in a direction normal to the defect. Since the magnetic particles are black, it is often necessary to paint the object under test white for direct viewing. Alternatively, fluorescent inks can be used for viewing under ultra-violet light.

An obvious limitation of this method is that it can be used only on ferromagnetic materials.

Dye Penetrant Method [6, 7]

A liquid penetrant coloured with a distinctive dye (normally red) is sprayed on to the test specimen. Time is allowed for the liquid to penetrate fully into any defects and then all traces of excess penetrant are removed from the surface. A developer consisting of an absorbent powder (normally white in colour) is then sprayed on. This causes any remaining penetrant, which has seeped into defects, to be drawn to the surface. Again, a fluorescent penetrant can be used for final viewing under ultra-violet light.

Penetrant testing is generally regarded as being less sensitive than the magnetic particle method although it is easier to apply with simple aerosol packaging. It is commonly used for quick checks in between weld runs or to ensure that the correct amount of back chipping or gouging has been achieved, and has the advantage of being applicable to any material.

Eddy Current Testing [7]

Eddy current testing is another method capable of detecting surface or near surface flaws but is not often used for testing welded joints because of the many variables that can affect the result, particularly metallurgical structure and surface profile. However, the method is sometimes used for the inspection of welded tube and recently claims have been made with regard to the in-service inspection of offshore structures [8].

The principle of the method is that eddy currents are induced into metals (ferrous and non-ferrous) whenever they are brought into an AC field, thus creating a secondary field which opposes the inducing field. The presence of discontinuities or variations in the material alters the eddy current, thus changing the apparent impedance of the inducing coil or of a detecting coil. The coils normally take the form of pencil-like probes scanned over the surface under test.

Depth Measurement

As already mentioned the inability to measure defect depth is a limitation of all the above methods. This is perhaps not a serious limitation when 'workmanship' acceptance criteria (see Chapter 6) are being used because the ability to detect and identify a planar defect would be sufficient. However, if fitness-for-purpose criteria are applied to either pre- or in-service inspection, a depth measurement of surface breaking defects is essential.

Attention has been paid recently to the development of methods capable of providing such a measurement. Two techniques appear to be particularly promising: one is the ultrasonic time-of-flight technique which is discussed later, the other is the AC potential drop (ACPD) technique. AC potential drop is based on the principle that a high frequency current tends to flow close to the surface (the 'skin effect') and therefore any potential drop measurement along the current path will be a function of the path length. Extension to the path length due to the presence of surface breaking defects will therefore increase the potential drop. Evaluations of this technique are currently being carried out and are due to be reported in the near future [9].

INTERNAL FLAWS

Most critical welded fabrications undergo examinations for internal flaws at some stage of construction. Increasing demand for joint integrity and design tightening has meant constant usage and development of these NDT methods. Of the two methods commonly used, radiography is the old-established technique used since 1917. Ultrasonic testing was introduced much later, in 1942. The principles, methods of application, capability and interpretation of these two methods are completely different in all respects.

Radiography [10]

Electromagnetic rays of the same nature as visible light, but of much shorter wavelength, have the ability to pass through solid objects. Absorption of the rays takes place according to the material density and presence of voids, inclusions, etc. Large slag inclusions and pores in weld metal absorb little radiation and the shadow image pattern

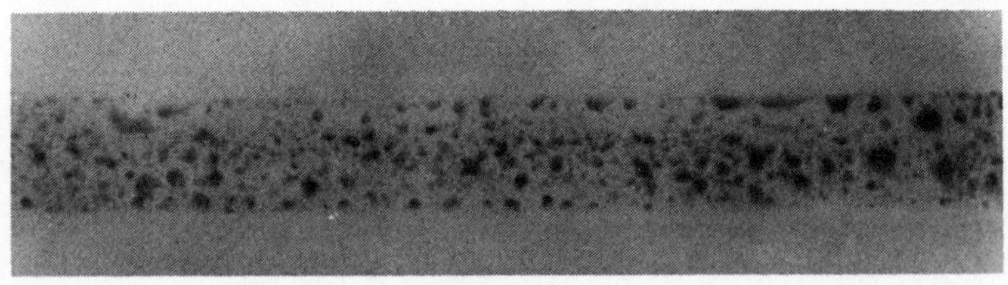

FIG. 1. Radiograph showing gross porosity in weld metal.

produced by such defects can be recorded on a photographic emulsion. An example of a radiograph showing gross porosity is shown in Fig. 1.

As well as requiring equipment for the production and control of the electromagnetic radiation, dark room facilities for the development of photographic films are required. Additionally, because of the health hazard imposed by radiation of this type, safety aspects are a prime consideration. Two distinct types of electromagnetic radiation are used for weld inspection, X-rays, and gamma-rays.

(1) *X-rays* (*Wavelength Range* 10^{-5} *to* 10^{-8} *mm*): X-rays are generated in a Coolidge tube by bombarding a dense metal target with fast-moving electrons accelerated from a heated filament. The higher the voltage across the tube the shorter the wavelength and hence the greater the energy and penetrative power of the X-rays. For welds in steel, tube potentials of 50–400 kV are used to inspect thicknesses of 5–100 mm. For thicker sections, the more penetrative gamma-rays are required. Although the smaller X-ray sets are transportable, some of the larger sets of 300 kV and above are bulky and have to be housed in heavily shielded bays to prevent escape of radiation.

(2) *Gamma-rays* (*Wavelength Range* 10^{-7} *to* 10^{-10} *mm*): Gamma-rays differ from X-rays in two fundamental ways. First, they are continuously emitted from radioactive isotopes and cannot be switched off and on. Second, they have a single wavelength, characteristic of the particular isotope used (line spectrum), whereas X-rays are composed of various wavelength levels (continuous spectrum). Isotopes commonly used for weld testing are iridium-192, cobalt-60 and caesium-137. These are used to test steel thicknesses of 60–120 mm. The equipment is easily transportable and is of low initial cost compared to the X-ray sets needed to test comparable thicknesses. Strict safety precautions must be adhered to involving the use of thick-walled, lead-lined containers.

Radiographic testing is excellent for the detection of three-dimensional defects such as slag inclusions and porosity. Planar defects, however, are sometimes impossible to detect since the change in specimen thickness (and hence absorption) is minimal. If the direction of the electromagnetic beam is parallel to a planar defect, however, the plane of maximum thickness change is examined and detection may be possible, see Fig. 2.

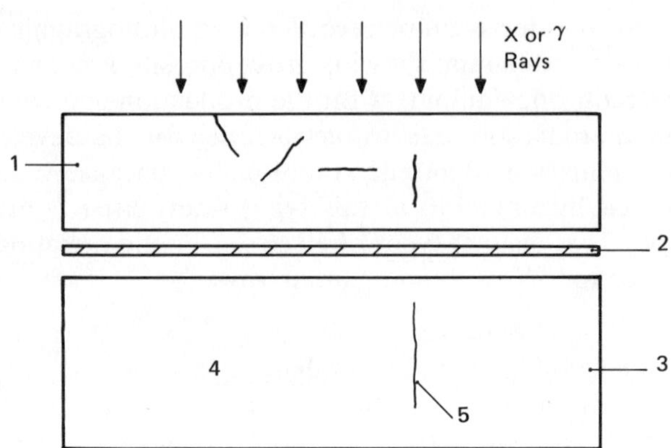

FIG. 2. Detection of planar defects by radiography (schematic). 1. Test specimen; 2. Film during exposure; 3. Developed film; 4. No images; 5. Image of vertical planar defect.

The sensitivity of radiography is governed by the definition and contrast of the resultant radiograph. These in turn are governed by the exposure conditions, photographic film, and development procedures; all of which are carefully controlled to achieve the required sensitivity. A measure of sensitivity (defined as the percentage thickness change detectable) is presented on the radiograph by means of an image quality indicator (IQI) [11]. This can be either a series of small wires or steps (of similar material to that under test) and is placed on the specimen at the maximum thickness of radiographic interest. Sensitivity is then measured from the smallest wire or step visible on the radiograph.

Interpretation of radiographs requires personnel with skill, experience, and a knowledge of what to look for in any particular circumstance. A dark room and viewing equipment with high intensity illuminators is necessary for critical interpretation. Small magnifying glasses are sometimes used if the film grain size is adequately small. A major advantage of the radiographic method is that the results can be easily stored for subsequent viewing and discussion by as many personnel as desired. However, problems can be caused by disagreements amongst radiographic interpreters.

One way of increasing radiographic sensitivity is to reduce the size of the radiation emitter (focal spot) [12]. This reduces the penumbral unsharpness resulting from the use of a focal spot of finite size.

Microfocus X-ray tubes are now available commercially with spot size down to 10 μm (c.f., 2–4 mm for conventional sets) and are applied in instances where high sensitivity and resolution are of paramount importance, for example, in the inspection of microcircuit interconnections.

Apart from insensitivity to planar defects, another limitation of radiography is the difficulty in measuring defect size in the through-thickness direction. Some work has been carried out on the use of densitometric methods [13], that is, measuring the radiographic density (darkness) of the defect image and comparing it with the density of sound areas. However, these methods are prone to error and can only be regarded as qualitative at present.

Display systems involving fluoroscopy can be used at the expense of sensitivity to speed up the radiographic testing cycle, for example, in the inspection of longitudinal welds in line pipe; and many additional aids such as image enhancement and processing (analogue and digital) are currently being studied [14].

As far as gamma-ray testing is concerned, a new low energy radioisotope source, ytterbium-169, has been shown to have major advantages over other sources for the inspection of thin-wall small-bore tubes [15].

Ultrasonic Testing [16]

Ultrasonic inspection of welded joints is based on the fact that a discontinuity within the material will reflect high frequency elastic vibrations propagated through the material. The pulse echo technique is used almost exclusively. Electrical pulses are generated in a test instrument and converted by a piezoelectric transducer into mechanical vibrations. The transducer normally takes the form of a hand-held probe scanned on the material under test using a suitable liquid couplant. Reflected pulses are received (generally by the same transducer) and converted back into electrical energy for display on a cathode ray oscilloscope (CRO) after being rectified and smoothed. Figure 3 shows a typical ultrasonic test being carried out.

All defect types are generally easily detected by the ultrasonic method. For the best detection capability of planar defects, the ultrasonic beam must be normal to the plane of the defect.

Two modes of wave motion are used for weld testing: compression and shear. Compression (or longitudinal) waves are caused by successive tension and compression of particles along the direction of propagation. Shear (or transverse) waves are caused by a shearing action of

FIG. 3. Manual ultrasonic test.

particles across the direction of propagation. Compression waves are introduced at right angles to the plate surface and are normally used for thickness gauging and to check for plate laminations or inclusions prior to weld testing. Where the weld has been ground smooth, a compression wave probe may be used in conjunction with shear wave probes to detect weld defects. Various angle shear wave probes are used so that there is the best chance of detection irrespective of defect orientation.

The test frequency is chosen on the basis of a trade-off between penetrative power (low frequency) and detection capability (high frequency). A frequency in the range 2–6 MHz is commonly used for weld testing. Sensitivity is controlled on the flaw detection equipment by a calibrated attenuator and a normally uncalibrated amplifier gain. Setting sensitivity is achieved by either setting the echo from a small machined defect (e.g. 1·5 mm diameter side drilled hole) to a known level; or by setting 'grass' (the name given to noise on the CRO timebase) to a stipulated level. Alternatively, sensitivity can be set by using the distance-gain-size (DGS) system [17] which allows the minimum detectable size of a perfect disc shaped reflector to be quoted.

Having detected a defect, the next stage would be to locate, size and characterise the defect to the extent required by the code of construction in force. Location is facilitated by calibrating the timebase of the CRO, reading off the position of the echo on the timebase, and then applying simple trigonometry. Location accuracy is generally good and typically within 5 % of the wall thickness.

Sizing defects by ultrasonics has been a major source of controversy for many years. In the USA and on the continent of Europe, it is common to use echo amplitude as a measure of defect severity [18] but this is an invalid assumption as amplitude is also governed by other factors, particularly defect orientation. In the UK, the most widely used sizing techniques for welds rely on defining the defect extremities by measuring the probe movement which gives a stipulated amplitude drop (20 dB for small defects). This requires more thorough calibration of probe characteristics than the amplitude techniques. However, probe movement techniques are also subject to errors. The magnitude of these errors has been assessed empirically in recent work and found to be significant (typically ±5 mm at the 95 % probability level) [2].

Characterisation of defects (i.e. predicting defect type) by conventional ultrasonics is also difficult. Identifying the complex source of a response from a single 'blip' on a CRO is obviously subjective and open to question. In some cases however, directionality of the source (i.e. detectable from one angle but not from another) can be used to indicate a planar defect.

The above limitations often create problems when attempting to interpret the results of an ultrasonic test to a given code particularly since the latter have tended to evolve from radiographic inspection standards. A great deal of research and development work has been carried out to solve some of the problems on the basis that ultrasonic testing is the only method which has the potential to provide all the necessary answers, and significant progress has been made as discussed below.

Other limitations of ultrasonic testing are: difficulties in testing materials with coarse grains, e.g. austenitic steel weld metal [19], due to severe disruption of the ultrasonic beam; and difficulties with welds in thin materials (e.g. less than 6 mm) where the volume being inspected is appreciably less than the ultrasonic beam diameter.

Attempts to assist the interpretation of ultrasonic tests have involved

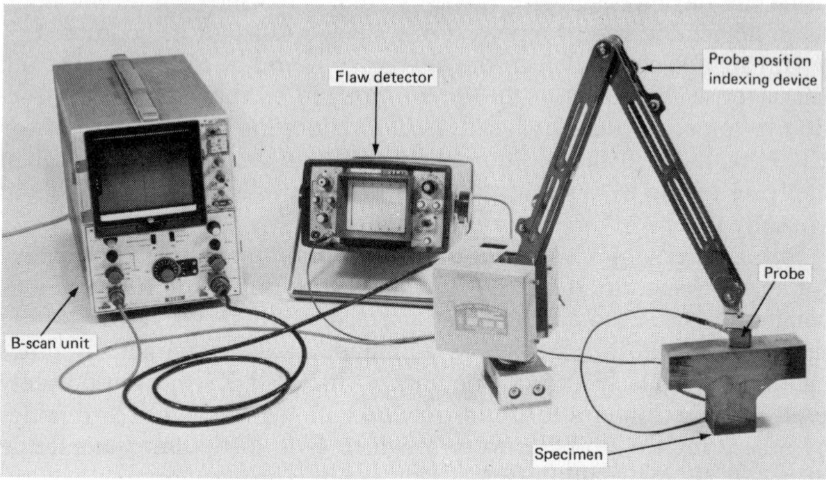

FIG. 4. Portable B-scan equipment.

the development of better display techniques. The ability to view the defect in its correct position in relation to the weld has clear advantages over the conventional amplitude/timebase (A-scan) display. One of the easiest ways of achieving cross sectional displays is with the portable B-scan system, see Fig. 4. This employs a conventional flaw detector and probe and the B-scan display is derived from echo data and probe position data. A typical display is shown in Fig. 5. Photographs of the display can be taken to provide a permanent record and a

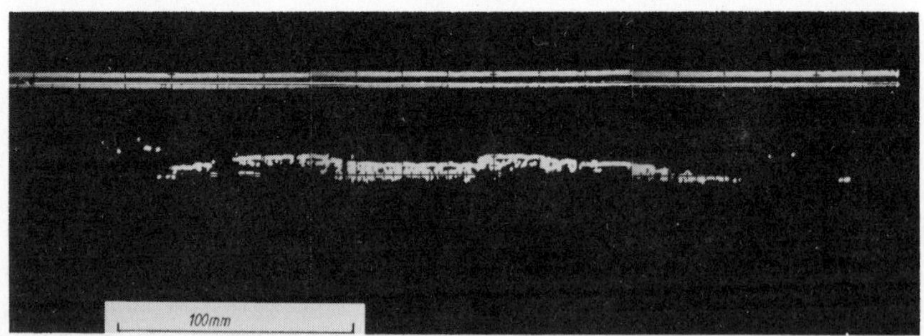

FIG. 5. Montage of B-scans showing longitudinal section of large weld metal crack and its position relative to upper surface.

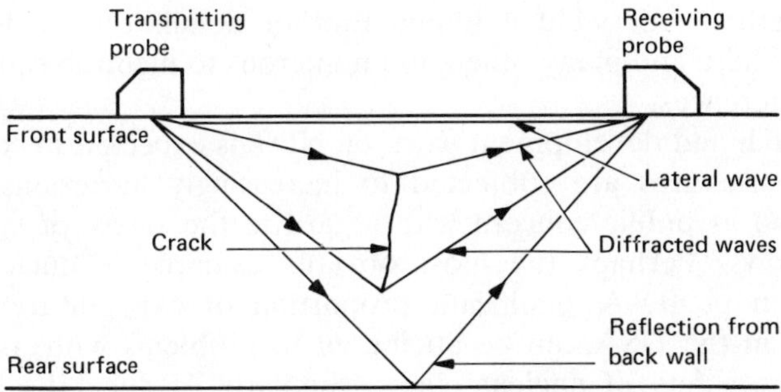

FIG. 6. Schematic arrangement of ultrasonic time-of-flight measurement technique.

variety of scan modes allows a comprehensive picture of the defect to be built up. The equipment has been thoroughly evaluated [2] and used to assist with defect assessment problems in many applications.

Digital techniques can be used to advantage in ultrasonic testing. The P-scan system [20] uses such techniques to provide amplitude data, cross-section maps, and horizontal section (C-scan) maps of defects all on the same display. Other variants of this type of equipment are available, for example 'Accuscan' [2] which gives more amplitude detail on the defect maps to assist characterisation and has recently been expanded to allow probe movement measurements to be made of defect through-thickness size [21].

The inherent inaccuracies in probe movement sizing have prompted workers to examine new techniques. The time-of-flight technique was originally developed for surface breaking defects [22] but has now undergone substantial modification to allow internal defects to be sized [2]. The principle is shown in Fig. 6. By measuring the arrival times of the separate diffracted waves from the defect extremities and knowing the velocity in the material and the separation of the transmitting and receiving probes, the vertical through-thickness dimension of the defect can be estimated. The technique has been shown to be substantially more accurate than probe movement techniques for measuring both planar and non-planar weld defects [2].

CONCLUDING REMARKS

It is hoped that this chapter has provided the reader with some guidance on principles of operation, application and current developments of

NDT methods for welded joints. Further details on the techniques discussed here and many others too numerous to mention can be found in the references.

Research and development work on NDT is expected to continue as welded structures are subjected to increasingly hazardous environments and as public concern with regard to the safety of large structures grows. Perhaps the most obvious example is nuclear power generation plant. A significant proportion of expenditure on NDT research in the USA can be attributed to problems with, or concern about, the safety of nuclear plants. Similarly, in the UK, the safety case for the pressurised water reactor system has provided the impetus for rigorous evaluations of NDT capabilities [3]. These efforts, on both sides of the Atlantic, are rewarding not only in solving the problems to which they are addressed, but also in the many spin-offs which find applications in a large number of other industries.

It cannot be claimed that NDT is now capable of providing all the answers, but developments in recent years have been significant in determining what can and cannot be achieved with both conventional and more specialised equipment. Digital techniques are still relatively new to the field of NDT and further advances in both radiography and ultrasonic testing can be expected as the full advantages of these techniques are realised.

REFERENCES

1. Anon. (1975). Major areas for research—ultrasonic testing, *Metal Construction,* **7**(11), 562.
2. Jessop, T. J., Mudge, P. J., Cameron, A. G. B., Charlesworth, J. P., Silk, M. G. Clare, A. B., Coffrey, J. M. and Bowker, K. J., 'Size Measurement and Characterisation of Weld Defects in Ferritic Steel by Ultrasonic Testing' Part 1 (March 1979). Non-planar Defects, Part 2 (October 1982). Planar Defects, Part 3 (October 1982). Metallurgical Features, Part 4 (January 1983). Complex Geometries and Practical Applications, The Welding Institute Report Series Publications.
3. 'A Symposium on the UKAEA Defect Detection Trials', October 1982, Warrington, England.
4. Jubb, J. E. M. (1981). Undercut or toe groove—the cinderella defect, *Metal Construction*, **13**(2), 94.
5. Lumb, R. F. and Winship, P. (1977). Magnetic particle crack detection, Part 1: Basic principles, limitations and parameter definitions, Part 2: Flux generation and potential applications to offshore structures, *Metal Construction,* **9** (8 & 9), 293, 331.

6. BS 4416:1969, Method for Penetrant Testing of Welded or Brazed Joints in Metals.
7. Blitz, J. G., King, W. G. and Rogers, D. G., *Electrical, Magnetic and Visual Methods of Testing Materials,* Butterworths, London, 1969.
8. Chisholm, J., Introducing the EMD MK III—A New Dimension in NDT, Offshore Oil International, 1st October 1981.
9. Webborn, T. J. C., The Unit Inspection Company, Swansea, personal communication.
10. The Welding Institute, *Handbook of Radiographic Apparatus and Techniques,* Cambridge, 1973.
11. BS 3971:1980, Specification for Image Quality Indicators for Industrial Radiography (including guidance on their use).
12. Boogerd, W. J., Van Zuylen, P. and Fontijn, L. A. (1976). An unconventional 150 kV microfocus X-ray equipment for NDT purposes, *British Journal of NDT,* **18**(6), 175.
13. Placious, R. C., Garrett, D. A., Kasen, M. B. and Berger, H. (1981). Dimensioning flaws in pipeline girth welds by radiographic methods, *Materials Evaluation,* **39**(8), 755.
14. Halmshaw, R., *Industrial Radiography; Theory and Practice,* Applied Science Publishers, London, 1982.
15. Pullen, D. and Hayward, P. (April 1979). Gamma radiography of welds in small diameter steel pipes using enriched ytterbium-169 sources, *British Journal of NDT,* **21**(4), 179.
16. The Welding Institute, *Handbook on the Ultrasonic Examination of Welds,* Cambridge, 1979.
17. BS 3923, Methods for Ultrasonic Examination of Welds, Part 1: 1978 Manual Examination of Fusion Welds in Ferritic Steels.
18. American Welding Society, AWS D1.1-80 Structural Welding Code—Steel.
19. Whitaker, J. S. and Jessop, T. J. (Nov. 1981). Ultrasonic detection and measurement of defects in stainless steel—a literature survey, *British Journal of NDT,* **23**(11), 293.
20. Nielsen, N. (March 1981). P-scan system for ultrasonic weld inspection, *British Journal of NDT,* **23**(3), 63.
21. Whittle, M. J., CEGB, NDT Applications Centre, Manchester, personal communication.
22. Silk, M. G. and Lidington, B. H. (March 1975). Defect sizing using an ultrasonic time delay approach, *British Journal of NDT,* **17**(2), 33–6.

APPENDIX—RELEVANT STANDARDS

British Standards

BS 3683: Part 2: 1963, Glossary of Terms Used in NDT—Magnetic Particle Flaw Detection.

BS 4069: 1966, Magnetic Flaw Detection Inks and Powders.

BS 4080: 1966, Methods for Non-destructive Testing of Steel Castings.
BS 4124: Part 2: 1968, Non-destructive Testing of Steel Forgings—Magnetic Flaw Detection.
BS 4397: 1969, Methods for Magnetic Particle Testing of Welds.
BS 6072: 1981, Methods for Magnetic Particle Flaw Detection.
BS 3683: Part 1: 1963, Glossary of Terms Used in NDT—Penetrant Flaw Detection.
BS 3889: Part 3A: 1965, Methods for Non-destructive Testing of Pipes and Tubes—Penetrant Testing of Ferrous Pipes and Tubes.
BS 4124: Part 3: 1968, Non-destructive Testing of Steel Forgings—Penetrant Flaw Detection.
BS 4416: 1969, Method for Penetrant Testing of Welded or Brazed Joints in Metals.
BS 4489: 1969, Method for Assessing Black Light Used in Non-destructive Testing.
BS 3683: Part 5: 1965, Glossary of Terms—Eddy Current Flaw Detection.
BS 3889: Part 2A: 1965, Eddy Current Testing of Ferrous Pipes on Tubes.
BS 3889: Part 2B: 1966, Eddy Current Testing of Non-ferrous Tubes.
BS 499: Part 3: 1965, Terminology of and Abbreviations for Fusion Weld Imperfections as Revealed by Radiography.
BS 2597: 1965, Glossary of Terms Used in Radiography.
BS 2600: 1973, General Recommendations for the Radiographic Examination of Fusion Welded Butt Joints in Steel.
BS 2633: 1966, Class 1 Arc Welding of Ferritic Steel Pipework for Carrying Fluids.
BS 2910: 1973, General Recommendations of the Radiographic Examination of Fusion Welded Circumferential Butt Joints in Steel Pipe.
BS 3385: 1961, Direct-reading Personal Dosemeters for X and Gamma-Radiation.
BS 3451: 1961, Testing Fusion Welds in Aluminium and Aluminium Alloys.
BS 3455: 1962, Glossary of Terms Used in Nuclear Science.
BS 3510: 1968, A Basic Symbol to Denote the Actual or Potential Presence of Ionising Radiation.
BS 3513: 1962, Gamma-radiography Sealed Sources.
BS 3683: Part 3: 1964, Glossary of Terms Used in Non-destructive Testing—Radiological Flaw Detection.

BS 3971: 1980, Image Quality Indicators for Radiography and Recommendations for Their Use.

BS 4097: 1966, Gamma-radiography Exposure Containers for Industrial Purposes, and Their Source Holders.

BS 4206: 1967, Methods of Testing Fusion Welds in Copper and Copper Alloys.

BS 2704: 1978, Calibration Blocks and Recommendations for Their Use in Ultrasonic Flaw Detection.

BS 3683: Part 4: 1965, Glossary of Terms—Ultrasonic Flaw Detection.

BS 3923: Part 1: 1978, Methods for Ultrasonic Examination of Welds—Manual Examination of Welds in Ferritic Steels.

BS 3923: Part 3: 1972, Methods for Ultrasonic Examination of Welds—Manual Examination of Nozzle Welds.

BS 4331: Part 1: 1978, Methods for Assessing the Performance Characteristics of Ultrasonic Flaw Detection Equipment-overall Performance: On-site Methods.

BS 4336: Part 1A: 1968, Methods of NDT of Plate Material—Ultrasonic Detection of Laminar Imperfection in Ferrous Wrought Plate.

American Standards

AWD1.1—1980, Structural Welding Code—Steel. American Welding Society.

ASTM Annual Book of Standards Part II, Metallurgy, Non-Destructive Testing, 1980.

ASME Code Section V Non-Destructive Examination.

West German Standards

DIN 54112, Non-destructive Testing, Films, Exposure Screens, Cassettes for X and Gamma-ray Radiographs, Dimensions.

DIN 54116 T1, Non-destructive Testing, Film Viewing, Viewing Conditions.

DIN 54116 T2, Non-destructive Testing, Viewing Films, Film Viewing Boxes.

DIN V 54119, Non-destructive Testing, Ultrasonic Testing, Definitions.

DIN E 54119, Non-destructive Testing, Ultrasonic Testing, Concepts.

DIN E 54124 T1, Non-destructive Testing, Control of Test Equipment with Ultrasonic Echo Instruments, Control at the Test Place.

DIN E 54126 T1, Non-destructive Testing, General Rules for Ultrasonic Testing, Requirements for Test Systems and Test Objects.

DIN E 54126 T2, Non-destructive Testing, General Rules for Ultrasonic Testing, Performance of Test.
DIN 54131 T1, Non-destructive Testing, Magnetising Equipment for Magnetic Particle Inspection–Stationary and Transportable Equipment Except Transportable Magnetic Yoke.
DIN 54132, Non-destructive Testing, Determining the Properties of Test Media for the Magnetic Particle Test.
DIN 54140 T1, Non-destructive Testing, Electromagnetic (Eddy Current Methods), Generalities.
DIN 54152 T1, Non-destructive Testing, Penetrant Testing, Execution.
DIN 54109 T1, Non-destructive Testing, Image Quality of Radiographs of Metallic Materials, Terms, Image Quality Indicators, Determinations of Image Quality Values.
DIN 54109 T2, Non-destructive Testing, Image Quality of X-ray and Gamma-ray Radiographs on Metallic Materials, Directives for the Classification of the Image Quality.
DIN 54111 T1, Non-destructive Testing, Testing of Welds of Metallic Materials by X- or Gamma-rays, Radiographic Techniques.
DIN 54123, Non-destructive Testing, Ultrasonic Methods of Testing Claddings Produced by Welding, Rolling and Explosion.

8

Codes and Standards Relevant to the Quality Assurance of Welded Constructions

J. H. Rogerson
Cranfield Institute of Technology, Bedford, UK

INTRODUCTION

In the quality assurance and quality control of any manufactured product it is necessary to have defined standards against which quality can be assessed. Such standards will include not only standards of performance of the completed article but also design standards, construction standards (e.g. for dimensional tolerances or methods of manufacture), inspection standards (standardised inspection and test methods) and standards for the materials of construction. Since welding is defined in quality assurance terms as a 'special process' it follows that the ability of the welders and the procedures they use must come under the scrutiny of the quality control and quality assurance systems. Therefore a further set of standards must be defined which relate to the welder's ability (welder approval standards) and the quality of a welding procedure (welding procedure approval standards).

The standards, codes and specifications which are particularly relevant to the quality control and quality assurance of welded fabrications can therefore be categorised as follows:

(1) Standards for consumables (e.g. BS 639, DIN 1913, AWS A5.1.78).
(2) Standards for welding procedure approval (e.g. BS 4870, ASME Section IX).
(3) Standards for welder approval (e.g. BS 4871, BS 4872, ASME Section IX, DIN 8560).
(4) Standards for the quality of a completed fabrication (e.g. application standards such as BS 5500, ASME Section VIII, AWS Structural Welding Code).

One further category of standards exists although there is only one well established relevant standard and that is:

(5) Quality assurance guidelines for welding operations (DIN 8563).

In this context the subtle difference between 'standards' and 'specifications' must be emphasised as the terms are frequently used indiscriminately. A standard is defined as 'a thing serving as basis of comparison', or 'weight or measure by which the accuracy of others is judged' whereas a specification is defined as 'detailed description of construction, workmanship, materials, etc., of work undertaken by an engineer' [1].

As well as there being generally agreed national and international standards relating to the construction of welded articles there are frequently more specific requirements relating to a particular product which are imposed (usually) by the client. Such requirements are the specifications which are often more rigorous and definitive than the relevant standards. A current example is the specification of welding consumables for the welding of primary structures for offshore oil and gas installations in the northern North Sea. Because of the concern about the possibility of fracture it is considered necessary that consumables should be capable of depositing weld metals of a very high fracture toughness at low temperatures. Such a specific quality requirement is not included in the available standards for welding consumables (e.g. BS 639 or AWS A51.78) so clients will specify fracture toughness levels for consumables to be used on their projects. The use of specifications as well as established standards is inevitable given the varied nature of welded products but badly devised specifications can cause trouble for fabricator and client alike. Standards, because of the wide range of experience that is likely to be consulted in their formulation, may be excessively conservative but are very rarely badly devised. It is therefore good practice to work as far as possible within the confines of established standards.

STANDARDS FOR WELDING CONSUMABLES

It must be recognised that a standard for a welding consumable, be it for an electrode, a MIG wire or a submerged arc flux cannot entirely define the performance characteristics of the consumable. More detailed and precise specifications may be able to do this for particular circumstances. A standard (or a specification) can only usefully define those attributes of a material which can be quantified and readily measured. In the case of a welding consumable many of the quantifiable attributes of interest are attributes of the deposited weld metal rather than the consumable itself so there is usually a requirement for 'standard' test welds as part of the standard for a consumable.

We can group the properties of a consumable which can be defined for the purposes of a standard into three categories:

(1) Dimensions (for example electrode size, wire diameter, spool size)
(2) Packaging (for example type of packaging for electrodes and fluxes, identification markings of type and batch number)
(3) Properties of the weld deposit (chemical composition, mechanical properties and soundness of the deposit)

Standards for dimensions are easily definable and must obviously be compatible with related standards (for example standards for wire and rod and standards for welding equipment dimensions and capacities). There is no real problem in this area and it is rare for welding consumables not to conform to dimensional standards and tolerances.

It is less easy to precisely define a packaging standard, particularly for a hydrogen controlled consumable, but, again, this is not normally a problem area in a quality assurance sense. The marking and identification of consumables has been the subject of much dispute [2] but this is more a problem of control of materials in a fabricator's stores and workshops than a problem of the standards themselves. For instance, however well the marking system for bags or tins of submerged arc flux is defined it has little relevance to the control of flux on the shop floor when excess flux is being recycled into the flux hoppers.

The contentious and difficult aspect of standards for consumables is the definition of performance characteristics. It is possible to write standards which define the mechanical properties of a weld deposit (usually the yield strength and sometimes the fracture toughness—the latter in terms of an impact test requirement) and the chemical composition, but more difficult to write standards which quantitatively define the soundness of the deposit. This very restricted set of requirements for performance standards clearly leaves much undefined about the properties of a consumable particularly as the performance characteristics are measured on a test weld which may not be representative of any of the welds in a particular structure.

This being the case it is necessary to view standards for welding consumables more as a way of classifying types of consumable and aiding the manufacturer's quality control system rather than as a way of defining performance standards. The corollary to this is that the welding procedure approval exercise takes over the role of defining the quality of the consumables because if the consumable used in the procedure test produces a weld which meets the performance requirements then this implies that the quality of the consumable is satisfactory in terms of its performance characteristics.

British Standards

If we consider three typical British Standards for consumables—BS 639 Covered Electrodes for the Manual Metal Arc Welding of Carbon and Carbon–Manganese Steels, BS 4165 Electrode Wires and Fluxes for the Submerged Arc Welding of Carbon Steel and Medium Tensile Steel and BS 2901 Filler Rods and Wires for Gas Shielded Arc Welding—we can see the limitations of such standards from a quality assurance point of view.

BS 639 defines an electrode in terms of mechanical property level, coating type (e.g. rutile, cellulosic, basic), deposit efficiency, electrical polarity and positional capability. The mechanical property levels are assessed by all weld metal tensile and impact tests and bend tests on standardised test welds. Hydrogen levels in hydrogen controlled electrodes are also assessed by a standardised method. These requirements have to be met initially and checked at least every 12 months by the appropriate tests to comply with the standard. It is also required that the manufacturers should have an adequate quality control system but the requirements for it are not defined.

BS 4165 has a similar approach except that a distinction is made between multirun and two run welds for mechanical property classification. In particular deposit compositions are not defined (except for maximum levels of sulphur and phosphorus).

BS 2901 defines the electrode properties in terms of the chemical composition of the wire and makes no attempt to define deposit properties either in terms of composition or mechanical properties. This is because the deposit properties are significantly affected by the choice of shielding gas and the welding process characteristics.

German Standards

These follow, in a technical sense and in their philosophy, very closely to the BSI standards—a typical example being DIN 1913

Covered Electrodes for Joint Welding of Unalloyed and Low Alloy Steels.

American Standards

The appropriate standards are those produced by the American Welding Society and, in a technical sense, are very similar to the above mentioned British Standards but with some important differences. There is an attempt to define the 'useability' of the consumable and some guidance is given on quality assurance in the procurement of consumables.

If we take as examples AWS A5.1.78 Specification for Carbon Steel Covered Arc Welding Electrodes and AWS A5.17.80 Specification for Carbon Steel Electrodes and Fluxes for Submerged Arc Welding it can be seen that consumables are classified according to mechanical properties, type of coating, position and type of current, i.e. very similar to the BSI classification. Weld deposit properties are also defined in a very similar way—a standardised test weld is made which is evaluated in terms of transverse tensile strength, all weld metal tensile strength, deposit chemical composition, impact strength and ability to pass a guided bend test (though not all of these tests are required for every consumable class). Extra tests though are required to assess the useability of the consumable in terms of its ability to deposit sound weld metal. This is done by the radiography of butt welds (assessed according to a defined quality standard) and the visual inspection and fracture testing of fillet welds. This type of test is a very valuable part of a standard for welding consumables in that it defines an acceptable level of 'useability'.

The previously mentioned standards (British, American and German) define the properties of consumables but they do not define precisely how often these properties should be checked (except that BS 639 requires periodic tests every 12 months) and this is necessary to control and assure the quality. Without a defined intensity and frequency of testing the fact that a consumable has once been shown to meet a particular consumable standard requirement is of little value to the user. American Welding Society A5.01.78 Filler Metal Procurement Guidelines goes a long way towards defining such matters. Its scope includes the following section: 'It is intended to provide a method for preparing those portions of a procurement document which consists of the following: (1) the filler metal classification, (2) the lot classification, (3) the intensity of testing . . .'. It follows good QA practice by asserting that the consumable manufacturer should have a QA programme and states that a consumable manufacturer certifies that his product meets the appropriate AWS classification if he puts a label on his product identifying it as such. The terminology used in consumable production (dry batch, dry blend, heat, lot, etc.) is defined and the intensity of testing is defined at different levels ranging from 'the manufacturer's standards' to 'all tests specified by the purchaser for each lot shipped'. This standard therefore provides a well defined base upon which the quality assurance of consumables can be established by agreement between purchaser and supplier.

STANDARDS FOR WELDING PROCEDURE APPROVAL

There are a large number of standards and specifications for procedure approval. As well as national and international standards classification societies, some large client organisations (The Ministry of Defence for example in the UK) establish their own standards for different products. This proliferation is undesirable and costly to the fabricator and ultimately to the client. Farrar [2] has estimated that a medium size fabricator could spend more than £100 000 (1978 prices) per annum on welding procedure qualification tests if he must satisfy the requirements of a number of standards and specifications.

The reason for welding procedure approval is to demonstrate that a given welding procedure, if followed, will produce welds of adequate quality. A welding procedure approval standard, therefore, provides a measure against which a particular procedure can be judged. All the standards are similar in that a test weld (or welds) is made which is representative of the procedure to be approved and is then subjected to a range of destructive and non-destructive tests to assess its quality. The requisite quality standard for a test weld is allied to the quality of welding required for the fabrication. The significant difference between different standards which cover the same type of fabrication and the one that has the most bearing on the cost of welding procedure approval lies in the range of welding procedures which are 'approved' or 'qualified' by the making of a given test weld to the requisite standard.

There is an important difference in philosophy between the American approach (typified by the ASME Boiler and Pressure Vessel Code Section IX, and the AWS Structural Welding Code) and the European approach (typified by BS 4870) to welding procedure approval standards. The welding procedure approval sections of these American codes are integral parts of codes which deal with all aspects of the relevant type of fabrication. For example, the ASME Boiler and Pressure Vessel Code covers design, materials, fabrication methods, inspection and quality assurance of welded pressure vessels as well as the qualification of welding procedures and welders. The AWS Structural Welding Code and API 1104 Standard for Welding Pipelines and Related Facilities similarly cover the whole range of quality related activities for the types of welded fabrication within their scope. The welding procedure approval requirements of these codes are therefore meant to apply only when *all* other code requirements apply. Consequently it is possible to limit the testing of a test weld to a relatively small number of tests and to permit some considerable latitude in the variations in welding procedure which are covered by a given test procedure. The European approach is in general different in that welding procedure standards (for example BS 4870 Approval Testing of Procedures) are designed for a wide range of products and circumstances. Consequently it is not possible to rely on other sections of a design and fabrication standard to limit the scope of the welding procedure standard. The amount of testing required on the test weld is therefore greater and the latitude in the variations of welding procedure covered by a given test procedure is less than in comparable American codes. Paradoxically, also, more flexibility and lack of precision is inherent in such a philosophy.

BS 4870: Specification for Approval Testing of Welding Procedures

This standard, being a British Standard, is accepted by all major interests in the UK and is the welding procedure approval standard called up in other British Standards (e.g. BS 5500 Unfired Fusion Welded Pressure Vessels). The standard lists the items to be recorded in the procedure test (e.g. base metal specification, edge preparation, run sequence, etc.) and stipulates the changes in procedure which invalidate an approval (i.e. the 'essential variables'). These are, briefly, changes in welding process, welding position, base metal thickness, range, electrode type, electrode size (subject to some exceptions), change in polarity, significant changes in preheat and post-weld heat treatment. Base metals are divided into groups and separate approvals are required for plate and pipe. Three test piece types are covered by the standard—a butt joint, a fillet joint, and a branch connection for pipes although 'special' test pieces which are more representative of the joint to be approved are also allowed.

All test welds must be subjected to visual, magnetic particle (or dye penetrant) testing, radiography and/or ultrasonic examination and the quality standard required to be achieved is high and equivalent to that demanded for production welds in BS 5500.

Butt joints require, in addition, macro-examination, a hardness survey, transverse tensile and all weld metal tensile tests, and bend tests whilst fillet joints require macro-examination, a hardness survey and (for single sided welds) a fracture test. Pipe branch connections require macro-examination and a hardness survey. A slightly ambiguous note suggests that more extensive testing (e.g. fracture testing) may be useful and advantageous.

This standard therefore requires a comprehensive testing scheme and is somewhat open ended in that special test pieces and/or extra tests may be included with the agreement of the contracting parties. This is the drawback of a standard which sets out to cover all circumstances. The fact that a fabricator has a procedure qualified to BS 4870 may therefore *not* mean that this qualification is acceptable to another client who also asks for procedure qualification to BS 4870. His interpretation of the requirements may be different even though it is expected that a procedure test witnessed and approved by one inspection authority to this standard should be acceptable, without further testing, to another inspection authority.

ASME Boiler and Pressure Vessel Code Section IX

Welding procedure approval to this standard can be equated with an approval to BS 4870 in that in each case procedures are being approved to the highest realistic standard—that demanded for the fabrication of pressure vessels. However, the requirements of ASME Section IX are very different from those of BS 4870.

The responsibility for qualifying procedures is held by the manufacturer (fabricator) and all procedures must be documented in a welding procedure specification (WPS) which lists the details of the procedure which are defined as either 'essential' or 'nonessential' variables. Each procedure must be qualified by the welding and testing of a test coupon which is recorded in a 'procedure qualification record' (PQR). This latter documents the essential variables of the welding procedure and any change in the essential variables requires a requalification of the procedure. Essential and nonessential variables are precisely defined but are, in many ways, less restrictive than in BS 4870. For example, a groove weld (butt weld) qualification test will also qualify procedures for fillet welds in all material thicknesses within the range of the other applicable essential variables for most base metal groups and combinations. Also, a qualification of a procedure in one welding position usually qualifies the procedure for use in all positions.

The test requirements of this code are less extensive than in BS 4870 and comprise, for groove welds, transverse tensile tests and bend tests and for fillet welds (where these are separately qualified) macro-examination of sections.

The apparently very inadequate checking of welding procedures which is specified by this code is acceptable because ASME Section IX assumes that *all* other relevant code requirements are being followed and that the manufacturer is authorised by ASME as a competent fabricator, i.e. the ASME code requirements presuppose that the fabricator who works to them has already been shown to be a competent and responsible fabricator of pressure vessels. This illustrates the danger of using ASME Section IX out of context and specifying its requirements as a convenient standard for approving welding procedures. Unfortunately ASME Section IX is frequently used in this way.

API 1104 Standard for Welding Pipelines and Related Facilities

This American code is widely used internationally as a guide for standards of weld quality in oil and gas pipelines. The standard requires that welding procedures be fully documented in terms of essential and nonessential variables, changes in the former of course require a requalification. Essential variables include, as expected, pipe materials, welding process, filler metal, welding position, flux, shielding gas and travel speed. Essential variables specific to pipe welding should be noted—changes in time lapse between root bead and second bead and changes in welding direction (vertical uphill to vertical downhill and vice versa).

The test requirements of this standard are transverse tensile (except for small diameter pipe), bend tests and nick break tests for butt welds and nick break tests for fillet welds.

Again, this limited degree of testing is generally acceptable because of the control exercised on the design and selection of materials and because a pipeline is a well defined and well understood 'structure'. In some special circumstances (e.g. very high yield strength steels, pipelines for arctic conditions or subsea pipelines) more extensive procedure testing is desirable particularly in terms of guaranteeing fracture properties or maximum HAZ hardnesses. In these cases API 1104 is not a sufficient standard for procedure approval.

AWS Structural Welding Code

This code covers a wide range of structures (excluding pressure vessels and pipelines) and is an appropriate code for buildings, bridges and tubular structures. It covers structures welded in carbon and low alloy steels and, like the ASME Boiler and Pressure Vessel Code, is a code complete in itself.

Groove (butt) weld procedures are qualified on the basis of bend tests and transverse tensile tests and fillet weld procedures are qualified on the basis of nick fracture and macro tests. Groove weld procedure qualification is also permitted on the basis of radiography instead of destructive tests. As in other standards there are essential and non-essential variables in a procedure, alteration in the former requiring requalification.

The distinctive feature about procedure approval according to this code is that procedures may be considered 'prequalified' (i.e. actual procedure tests are not required) provided that they meet certain criteria. The convenience and economic value of this to the fabricator and, ultimately to the client is obvious.

These criteria are that the welding process is MMA, submerged arc, MIG/MAG or flux core, the joint design is limited to those recommended in the code, the quality of workmanship (e.g. fit-up, distortion) meets the defined standards and the 'technique' also meets the defined standards (e.g. filler metal type, minimum preheat and interpass temperatures).

The fact that this method of procedure qualification is satisfactory further emphasises the philosophy that formal procedure approval testing is only one of many factors which make up a quality control and quality assurance system for welded fabrications.

STANDARDS FOR WELDER APPROVAL

Standards and specifications for welder approval (for which there are probably a greater number than there are for procedure approval) are superficially very similar to those for welding procedure approval but the aim is quite different. In this case the aim is to demonstrate that the welder is capable of producing a weld of acceptable quality provided he is given a suitable procedure to work to. Therefore in assessing the quality of a test weld any defects which do not relate to the welder (e.g. defects arising from faulty material or an inadequate procedure) should be discounted. Also the essential and nonessential variables are likely to be different even under the same code for welder approval and for procedure approval. A further point is that in most situations far more welder approvals will be needed than procedure approvals so economic necessity requires that welder approval testing should rely as far as possible on simple, quick (and therefore cheap) test methods. Finally, since a person's skill is being assessed, any standard must include a provision for ensuring that a welder, once qualified, does not automatically remain a qualified welder unless he is maintaining his skill. Most standards therefore either require a welder to be requalified at regular time intervals or else require a welder to be continuously employed on an appropriate type of welding and producing welds to a satisfactory standard to maintain his qualification. This latter approach is the more logical of the two and is the one most frequently adopted. It has the disadvantage, however, from the quality assurance auditor's point of view in being more difficult to 'police'. The main American and British codes have many similarities in that they tend to be tied to relevant procedure qualification standards. Some others (particularly DIN 8560 and 8561) seem to come from a different philosophy in that they try to define a welder's skill in a more general and wider sense.

BS 4871 Approval Testing of Welders Working to Approved Welding Procedures

This is the counterpart document to BS 4870 (see previous section). The test weld requirements are the same as for BS 4870 but the inspection and testing of the test weld is much more limited. Assessment can be either by non-destructive testing (supplemented by macro-examination or bend testing in certain specified situations) or by destructive testing (bend tests for butt welds and fracture tests for fillet welds). The quality standards demanded are equivalent to those in the counterpart standard for procedure approval (BS 4870). The essential variables are primarily position and base metal thickness range. Changes in procedure such as preheat or post-weld heat treatment for example do not require a welder requalification and there are also clauses which permit repeat tests to be taken if a test piece fails to meet the required quality standard as a result of metallurgical or extraneous causes not attributed to the welder's workmanship. Also, if the welder realises the test weld is likely to fail he is permitted to withhold its submission and prepare a second test weld. These differences between 4871 and 4870 are a recognition that it is the welder's skill and his ability to make judgements on his own work which are being assessed.

A welder, once qualified, remains qualified provided he can be shown to have been reasonably continuously employed on work of the appropriate quality and that his workmanship has been satisfactory.

BS 4872 Approval Testing of Welders When Welding Procedure Approval is Not Required

This standard is designed to cater for the very large category of welding where there is no mandatory or contractual requirement for the welding procedure to be formally approved but where a 'good' standard of welding is required. The quality of welding in such cases is very dependent upon the skill of the welder and BS 4872 defines an approval scheme for ensuring that welders in a wide range of industries can be qualified to a comparable standard.

A wide range of test pieces is specified (butt and fillet welds in sheet, single sided butt welds in plate with and without backing, double sided butt welds in plate, fillet welds in plate and butt and fillet welds in pipe) and assessment is on the basis of simple destructive tests (macro, bend and nick fracture tests) with a required quality standard somewhat lower than in BS 4871. Other than thickness range, position and joint type (which are essential variables) the essential and non-essential variables are not precisely defined because the very wide applicability of this standard makes it impossible. Such decisions must be made by the engineer (or inspecting authority) according to the particular circumstances. This lack of precision is a disadvantage because it may lead to incorrect and inappropriate interpretation of the standard but this is a minor point compared with the considerable advantage of having a sensible welder approval standard for the very large amount of welding which is not covered by the construction standards for pressure vessels, pipework, storage tanks and other high integrity components.

ASME Boiler and Pressure Vessel Code Section IX

A distinction is made between 'welders' and 'welding operators' and the qualification requirements are slightly different. Qualification in either case requires the welding of a test piece in accordance with a qualified welding procedure, with the option that for a welding operator the 3 ft length of his first production weld may be taken as the test piece.

The tests required of the test weld for welders are bend tests (groove welds) and macro and fracture tests (fillet welds) although for most processes and material groups the test weld for a groove weld qualification can be examined by radiography instead. In such a case his first production weld can be taken as the test piece. In all cases a welding operator may be assessed on the basis of radiography. The standard of quality required in the test piece is equivalent to that required for procedure qualification and, in fact, a welder or operator who makes the weld which qualifies a particular procedure is automatically qualified to use that procedure. The essential and non-essential variables as always in this code are precisely defined but are different from those for procedure qualification, the most important difference being that welding position is an essential variable for performance testing. In most cases a groove weld qualification will cover the welding of fillet welds but not vice versa.

The continuity of the approval is governed by similar rules to those in BS 4871 and 4872; the approval remains valid provided the welder or operator is essentially continuously employed on the appropriate quality of work.

API 1104 Standard for Welding Pipelines and Related Facilities

The welder qualification requirements fall into two parts—'single qualification' where the test weld is a butt weld (each position being an essential variable) or 'multiple qualification' where a butt weld in pipe of diameter and wall thickness equal to at least $6\frac{5}{8}$ in (168·3 mm) and $\frac{1}{4}$ in (6·35 mm) respectively is followed by the fitting and welding of a full size branch connection. If this multiple qualification is obtained then there are naturally fewer 'essential variables' which limit the welder's scope. As in the case of ASME Section IX, weld quality is evaluated by tensile, bend and nick fracture tests or, as an alternative, by radiography. There are no rigidly defined limits to the continued validity of a qualification except that it is stated that welder 'may be required' to requalify if there is a question about his ability.

AWS Structural Welding Code

The rules for welder qualification follow the same style as those in ASME Section IX except that the quality requirements for the test welds are lower and the latitude on variables is much greater. For example, a welder qualification on one steel type covered by the code covers the welding of all steels covered by the code. Welding process is an essential variable but a qualification with one electrode and shielding medium covers another electrode and shielding medium provided the process is the same.

Although the standard of attainment required to qualify according to this code is lower than that needed to qualify according to ASME Section IX it is probably higher than that required to qualify on an equivalent joint type according to BS 4872.

DIN 8560 Testing of Welders for Welding Steel and DIN 8561 Testing of Welders for Welding Non-ferrous Metals

These two German standards are, of course, similar in their approach and can be discussed together. They illustrate a very different philosophy of welder qualification than that followed in British and American codes. The British and American approach is to strictly limit qualification to a restricted range of joints and welding process variables usually in conjunction with a corresponding procedure (e.g. BS 4870 and 4871). This leads inevitably to a large number of qualification tests being carried out and a given welder often possessing a number of different qualifications. The DIN approach views a welder qualification more as a certificate the possession of which defines a person as a skilled craftsman. To qualify under DIN 8560 or 8561 a welder must first undergo a recognised training scheme (such as a DVS course) and as well as demonstrating his practical skills in a way (and to a level) very similar to that required in BS 4871 and ASME Section IX successfully pass a test of his theoretical knowledge of welding technology. This theoretical knowledge comprises safety precautions, operation of the equipment, preparation for welding, welding terms and symbols, knowledge of materials and of factors which govern the quality of a weld. All testing and examination must be carried out by an approved test centre (the normally expected method) or by a welding engineer recognised by the company. This approach, although presumably involving significant administration costs, is excellent from a quality assurance point of view in that nationally defined and administered standards for welder qualification lead to a more consistent standard of approval.

STANDARDS FOR THE QUALITY OF A COMPLETED FABRICATION

When we define the quality standard for a welded component we define it in terms of a defect level (type, size and number of defects) and sometimes, also, in terms of the mechanical properties of the weldment. In either case the definition of a method, or methods, of inspection together with the extent of inspection is necessary to establish the quality standard required. As stated in Chapter 5 there is very little available data on expected defect levels in welded fabrications in terms of type, size and, particularly, distribution. There is also very little data on the expected distribution of mechanical properties (tensile strength, toughness) along a weld made to a given procedure. This lack of data makes the setting of quality standards and the associated inspection requirements difficult with the consequence that arbitrary standards, which are sometimes difficult to justify logically, are often found. A further important generalisation which must be appreciated is that the higher the integrity required of the welded structure the greater will be the precision with which the quality standard (and associated inspection methods) will be defined.

Pressure vessel codes define the required weld quality standard very precisely and quantitatively whereas construction codes for 'general structures' (bridges, ships, buildings, for example) define quality standards in terms such as 'welds are to be sound, uniform and substantially free from slag inclusions and porosity'. This imprecision in the latter type of standard can lead to disputes about what is an acceptable quality level but, perhaps more importantly, results in the assurance of quality becoming more reliant on the design, material selection and welder and welding procedure approval. This means that the final inspection of the structure becomes relatively less important as a means of guaranteeing structural integrity.

Pressure Vessel Standards

It is not intended to discuss in detail the differing requirements of the major internationally used pressure vessel codes. In all cases the defined acceptance standards for weld quality are based on experience and judgement of what minimum quality standard a competent fabricator could be expected to consistently achieve. In only one of them (BS 5500) is there the possibility of defining a standard which is more logically related to the design and service environment of a particular vessel. Clauses 5.7.3.2. and 5.7.3.3. state that 'when acceptance levels different from those given in Table 5.7. have been established for a particular application and are suitably documented, they may be adopted by specific agreement' and 'particular defects in excess of those permitted in Table 5.7. may be accepted by specific agreement between the purchaser, the manufacturer and the inspecting authority after due consideration of material, stress and environmental factors'.

Normally however, weld quality for pressure vessels is defined in terms of acceptable defect levels which are disclosed by non-destructive testing. Although ultrasonic testing is widely used in the fabricating industry the quality standards are still derived on the basis of radiographic examination. Cracks and other planar defects such as lack of fusion, lack of penetration, aligned porosity are always unacceptable (except for the BS 5500 possibilities mentioned above), but there is a tolerance limit on volumetric defects such as porosity and slag inclusions and shape defects such as undercut. There are differences between the pressure vessel codes in terms of the acceptable level of such defects and obviously these differences are very important in a contractual sense. In practice the differences are insignificant from an engineering point of view and only arise because different committees have defined an arbitrary standard in slightly different ways.

The extent of non-destructive testing varies in that 100 % testing of welds is normally called for but for vessels of lower construction categories (lower stresses, simpler materials, less hazardous situations) only sampling inspection is called for. Although the results of such sampling inspection are judged against the same defect acceptance standards the minimum acceptable quality level will be lower in this case. The arbitrary and illogical nature of such sampling inspection schemes has been discussed elsewhere [4] but if they are considered as cost effective ways of ensuring that a certain but perhaps not closely defined quality level is maintained then they have some value. The fact which must be appreciated is that a sampling inspection scheme deals in probabilities not certainties and therefore with such an inspection scheme there is a significant probability that serious defects such as cracks will not be found. This situation must be catered for by either greatly reducing the risks of such defects occurring (procedure qualification and process control) or designing the structure such that its integrity is not impaired by their presence (e.g. ensuring that materials have a high fracture toughness or design stresses are low). This latter approach is the one which is being followed, of course, when sampling NDT is applied to pressure vessels.

Pipelines and Piping Systems

Quality standards for welds in pipelines and pipe systems can be compared to those for pressure vessels because in each case a fluid is being contained under high pressure and a similar degree of integrity is required. Furthermore, in many cases piping systems will be associated with pressure vessels in process plant and similar quality levels are obviously appropriate and are in fact defined in the various standards (e.g. API 1104 and BS 2633 Class I Arc Welding of Ferritic Steel Pipework for Carrying Fluids).

The various pipe welding standards have been written on the basis of manual metal arc welding, inspection by radiography and, particularly in the case of API 1104, the use of conventional pipe line steels. That is not to say that the standards do not cover other processes and materials as of course they do. The important consequence is that sometimes these standards are inappropriate (particularly if high yield steels or gas shielded arc welding processes are used) as they do not take into sufficient account fracture properties or the type of defects which may occur. This is an area where specification for quality standards and testing methods which are different from the available standards frequently need to be defined.

General Structures

The various standards which cover structures such as bridges, ships, buildings (examples are BS 5400, AWS Structural Welding Code, Classification Society Rules) do not define weld quality standards in a quantitative and therefore unambiguous way, that is if they attempt to define weld quality standards at all.

The difficulties in producing such definitions are, of course, great because 'general structures' covers a wide range of constructions which will require to be built to an equally wide range of quality levels. The best that standards can do in this area is to categorise structures into 'quality bands' and to define who should establish the quality standard and associated inspection level and at what stage in negotiations. The AWS structural welding code is perhaps a good model in this respect.

One further point about the AWS structural code which is very relevant in a quality assurance sense is that it very carefully discriminates between fabrication/erection inspection and testing which is the responsibility of the manufacturer and verification inspection and testing which is the responsibility of the owner. Inspection must also be carried out by AWS certified welding inspectors (i.e. to AWS QCI). The definition of responsibilities for quality related activities and the guarantee that these responsibilities are discharged by suitably competent people are fundamental principles of any quality assurance programme.

QUALITY ASSURANCE GUIDELINES FOR WELDING OPERATIONS

Apart from the sections of the general quality assurance system requirements which cover 'special processes' there is little in the way of standards which specifically set out to define the quality assurance requirements for welded structures. One document which does, however, is DIN 8563 Ensuring the Quality of Welding Operations. This consists of two parts. Part I summarises the general principles to be considered—description and certification of fabricators, technical equipment, personnel, design, manufacturing control and testing and is intended to provide guidelines for use when drawing up contractual agreements. All the essential QA/QC elements are there relative to welding operations but presented in a slightly different way than is normally found in British and US quality assurance standards. Part II is, in fact, the most valuable part in that it defines and standardises the QA requirements of a welding fabricator in such a way that it can be used for the assessment and certification of fabricators. Great emphasis is quite rightly laid on welding personnel and their qualifications—welding engineers, welding technicians as well as welders. It is implied, again quite rightly, that welding engineers and technicians should have formal qualifications and since this is a German standard such qualifications are defined in terms of DVS qualifications. This again is a fundamental aspect of a quality assurance programme—the responsibilities and competence of personnel must be appropriately defined.

CONCLUDING REMARKS

Standards are a vital part of any quality assurance activity because they provide the technical guidelines against which designs, material properties, construction standards and inspection methods are specified.

Traditionally, the welding fabrication industry has worked on a 'build then inspect' philosophy and only recently has moved towards a quality assurance philosophy. The technical standards related to welding fabrication inevitably reflect this traditional view and also reflect the great reliance placed upon the craft skill of the welder and the difficulty of defining important aspects of 'weldability'. This means for example that standards concerned with welder qualification and welding procedure qualification are well developed and very specific whereas standards concerned with welding consumables are less definitive. From a quality assurance point of view the greatest omission in many standards (notable exceptions being some of the American and German ones) is the clear definition of responsibilities and qualification requirements for those who discharge these responsibilities. It can be argued that such matters are not the concern of technical standards but it is in practice impossible to consider the technical requirements of a standard without considering also the manner in which these requirements are imposed and controlled. This is the core of quality assurance and quality control practices.

REFERENCES

1. *The Oxford English Dictionary*. Oxford University Press, Oxford.
2. Quality Control and Non-Destructive Testing in Welding, discussion in *Proc. Conf.*, The Welding Institute, 1974.
3. Farrar, J. C. M., Weld Procedure Qualification—The Costs and Benefits, Seminar Welding Approvals—a Testing Time for all, The Welding Institute Sheffield and East Midlands Branches, April, 1979.
4. Rogerson, J. H., The Implications of Sampling Inspection for the Quality Control of Welded Fabrications. IIW Public Session Quality Assurance in Welded Construction, Estoril, 1980.

Index